$1.25
look
#174

RADIO WAVE SCATTERING IN THE INTERSTELLAR MEDIUM

AIP CONFERENCE PROCEEDINGS 174

RITA G. LERNER
SERIES EDITOR

RADIO WAVE SCATTERING IN THE INTERSTELLAR MEDIUM

SAN DIEGO, CA 1988

EDITORS:

JAMES M. CORDES
CORNELL UNIVERSITY

BARNEY J. RICKETT
UNIVERSITY OF CALIFORNIA,
SAN DIEGO

DONALD C. BACKER
UNIVERSITY OF CALIFORNIA,
BERKELEY

AMERICAN INSTITUTE OF PHYSICS   NEW YORK 1988

Authorization to photocopy items for internal or personal use, beyond the free copying permitted under the 1978 US Copyright Law (see statement below), is granted by the American Institute of Physics for users registered with the Copyright Clearance Center (CCC) Transactional Reporting Service, provided that the base fee of $3.00 per copy is paid directly to CCC, 27 Congress St., Salem, MA 01970. For those organizations that have been granted a photocopy license by CCC, a separate system of payment has been arranged. The fee code for users of the Transactional Reporting Service is: 0094-243X/87 $3.00.

Copyright 1988 American Institute of Physics.

Individual readers of this volume and non-profit libraries, acting for them, are permitted to make fair use of the material in it, such as copying an article for use in teaching or research. Permission is granted to quote from this volume in scientific work with the customary acknowledgment of the source. To reprint a figure, table or other excerpt requires the consent of one of the original authors and notification to AIP. Republication or systematic or multiple reproduction of any material in this volume is permitted only under license from AIP. Address inquiries to Series Editor, AIP Conference Proceedings, AIP, 335 E. 45th St., New York, NY 10017.

L.C. Catalog Card No. 88-72092
ISBN 0-88318-374-9
DOE CONF-880191

Printed in the United States of America.

CONTENTS

Preface ix

I. SCINTILLATION PHENOMENA AND THEORY

Introduction to the Observables in Interstellar Radiowave Propagation 2
    B.J. Rickett
From Scintillation Observations to a Model of the ISM—
The Inverse Problem 17
    R. Narayan
Observations of Diffractive Interstellar Scintillation Phenomena 32
    S.R. Spangler

II. PHYSICS OF SCATTERING MEDIA

Cosmic Rays and the Physics of Interstellar Turbulence 48
    J. R. Jokipii
Turbulent Magnetohydrodynamic Density Fluctuations 60
    D. Montgomery
Interstellar Electron Density Fluctuations Due to Cosmic-Ray
Acceleration at Supernova Remnant Shock Waves 61
    C. E. Max, A. Zachary, and J. Arons
Shock-Associated MHD Waves: A Model for Interstellar Density
Fluctuations 66
    S. R. Spangler
Hydromagnetic Wave Heating of Low Density Interstellar Gas 70
    E. G. Zweibel, K. M. Ferriere, and J. M. Shull
Oceanic and Interstellar Fossil Turbulence 74
    C. H. Gibson

III. TURBULENCE IN THE SOLAR WIND

Radio Observations of Plasma Irregularities in the Solar Wind 82
    A. Hewish
The Solar Wind Turbulence Spectrum Near the Sun 87
    W. A. Coles and J. K. Harmon
Spatial Coherence in a Scatter-Broadened Image—Observations of 3C279
Close to the Sun 92
    T. J. Cornwell, K. R. Anantharamaiah, and R. Narayan
Interferometric Phase Scintillation in the Interplanetary Medium 97
    B. Dennison, R. S. Simon, S. Ananthakrishnan, and R. L. Fiedler

IV. DIFFRACTION PHENOMENA AND THE INTERSTELLAR ELECTRON DENSITY POWER SPECTRUM

VLBI Observations of the Scattering Disk of Pulsar 1933+16 106
    C. R. Gwinn, J. M. Cordes, N. H. Bartel, A. Wolszczan, and R. Mutel

Interstellar Scattering of SgrA* 111
  D. C. Backer
Angular Broadening Measurements of the Sources
1849+005 and 2013+370 117
  S. R. Spangler and J. M. Cordes
VLBI Measurements of Interstellar Turbulence Spectra 122
  R. L. Mutel and J. F. Lestrade
Interstellar Scattering of Radiation from H_2O Masers in W49 and Sgr B2 129
  C. R. Gwinn, J. M. Moran, and M. J. Reid
Interstellar Electron Density and Magnetic Field Irregularities on
0.001 to 100 Parsec Scales 134
  J. H. Simonetti and J. M. Cordes

V. REFRACTION PHENOMENA AND LARGE SCALE STRUCTURE

Refractive Effects in Pulsar Dynamic Spectra 140
  Y. Gupta, B. Rickett, and A. Lyne
Refractive Scintillation, Caustics, and Interstellar Interferometry of the
Pulsar PSR 1133+16 145
  A. Wolszczan, J. E. Bartlett, and J. M. Cordes
Extreme Scattering Events 150
  R. Fiedler, R. Simon, K. Johnston, B. Dennison, and A. Hewish
Properties of Large Dim Refractors 156
  R. W. Romani
Refractive Scintillation of Extragalactic Radio Sources 163
  W. A. Coles
Effects of Turbulent Interstellar Clouds on Refractive Scintillation 169
  R. G. Frehlich
Refraction from Interstellar Shocks 174
  A. W. Clegg, D. F. Chernoff, and J. M. Cordes

VI. THE GALACTIC DISTRIBUTION OF SCATTERING MATERIAL

Galactic Distribution of Electron Density Turbulence 180
  J. M. Cordes, S. R. Spangler, J. M. Weisberg, and T. R. Clifton
The Origin of Scattering in the Inner Galaxy 185
  K. R. Anantharamaiah and R. Narayan
VLBI Angular Broadening Measurements in the Cygnus Region 190
  A. L. Fey, S. R. Spangler, and R. L. Mutel
Preliminary Results from a 7 Station VLBI Survey of OH Masers in the
Galactic Plane 195
  P. J. Diamond, A. Martinson, B. Dennison, R. S. Booth, and A. Winnberg

VII. IMAGING TECHNIQUES AND PULSAR STUDIES

The Shape of a Scattered Image-Theory and Numerical Simulations 200
  J. Goodman and R. Narayan
Computer Modeling of Interstellar Scattering Effects on Pulsar Timing 205
  R. S. Foster and J. M. Cordes

Interstellar Interferometry 212
    J. M. Cordes and A. Wolszczan
Interstellar Scintillations of Binary Pulsars 217
    R. J. Dewey, J. M. Cordes, A. Wolszczan, and J. M. Weisberg
Low Frequency VLBI 222
    D. L. Jones

VII. AGENDAS OF PREVIOUS MEETINGS ON INTERSTELLAR SCATTERING

Interstellar Scattering of Radio Waves 228
    D. C. Backer and B. J. Rickett
Workshop on Interstellar Propagation of Radio Waves 229
    D. C. Backer

Author Index 230

Preface

Recent work on the scattering of radio waves in the interstellar medium has shown stimulating progress in the areas of (1) observational techniques and phenomena; (2) statistical optics and wave propagation; and (3) the physics of ionized turbulence in the interstellar medium. It was decided that a workshop followed by rapid publication of results would lead to cross fertilization of ideas in these subjects. A workshop, 'Radio Wave Scattering in the Interstellar Medium,' was held on 18–19 January 1988 at the University of California at San Diego. The organizers were Roger Blandford, Don Backer, Jim Cordes, and Barney Rickett.

Interstellar scattering is produced by refractive index variations in ionized regions of the interstellar medium. To a good approximation, these follow the cold plasma dispersion law so that the rich variety of observable effects directly probe electron density variations. These effects include angular broadening and wandering of radio source images; intensity scintillations in time and frequency; and temporal broadening and time-of-arrival variations of pulses from radio pulsars. Current interest includes situations where scattering is a source of noise that obscures signals of interest (as in the timing of radio pulsars) to those where the interstellar effects are the desired signals (used to study the interstellar medium) to the hybrid case where multiple imaging events caused by the medium allow unprecedented constraints on the structure in the emitting sources themselves (neutron star magnetospheres).

The main interest of the meeting was in the study of the interstellar medium. In interstellar scattering contexts, the relevant length scales (10^9 to 10^{15} cm) are orders of magnitude smaller than can be resolved by other astronomical techniques. The picture that has emerged over the last few years is that the interstellar microstructure is probably driven by high-energy phenomena—shocks from stellar winds and supernovae in particular. These same regions are also probable sites of cosmic-ray acceleration. The radio observations may therefore yield further insights on cosmic ray acceleration and propagation.

This volume contains the written versions of review talks, contributed talks, and posters that were presented at the meeting. The organization is topical and does not follow the order in which talks were given at the workshop.

The first chapter contains reviews on the interpretation of interstellar scattering observations. In the last few years, workers have categorized the observed phenomena into those resulting from 'refraction' through large scale irregularities in the interstellar medium and those from 'diffraction' by small scale turbules. Observables are generally expressible as functionals over the distribution of all turbules. Inversion of observables into constraints on the distribution is complex. The reviews convey this complexity yet demonstrate that the wave propagation physics is relatively well understood while the physics of the interstellar medium (and other scattering media) is not.

Chapter Two contains discussions on the physics of turbulence and density fluctuations and, in the case of the interstellar medium, their possible relationship to high-energy phenomena such as supernova shocks and cosmic-ray acceleration and propagation. For those who work predominantly in the observational and wave propagation areas, it is exciting to learn about the relevance of these areas to the global theories of the interstellar medium that have emerged over the last two decades.

Chapter Three is devoted to turbulence in the interplanetary medium (IPM). The IPM is interesting in itself, but in this context it is also viewed as a laboratory for studying turbulence such as may also exist in the interstellar medium. The ability to make *in situ* measurements is extremely important for testing theories of processes that may occur in both the IPM and the interstellar medium.

Chapter Four assembles observations that lead to constraints on the wavenumber spectrum of interstellar electron density variations. Most of these are interferometric measurements of the two (spatial) point correlation of the electric field (the visibility function). One concerns image wandering and another discusses Faraday rotation variations. When cast in terms of a power law model for the wavenumber spectrum, all the results appear to favor a spectrum that is 'shallow' and causes wave propagation to be dominated by diffractive, rather than refractive effects.

Whereas Chapter Four presents results that are in general agreement about the shape of the wavenumber spectrum, Chapter Five concerns refraction phenomena that suggest that the form of the spectrum is perhaps not as stable as is implied. Topics in this chapter include the influence of refracting entities on diffractive scintillations; the theory of 'statistical' refractive scintillations such as would arise from a distribution of length scales; and observations and theory of lensing events from discrete entities in the interstellar medium. The evidence for discrete entities has grown quite strongly in the last two years but their relationship to statistical variations in the interstellar medium is not yet clear.

Discussions about the distribution of scattering material in the Galaxy are contained in Chapter Six. The papers in this section emphasize the extraordinary variation of the level of scattering and suggest possible provenences for the material in terms of the various density and temperature phases of the interstellar medium.

Chapter Seven presents new results in the theory of images produced by the turbulent interstellar medium; the influence of scattering on the precision of time-of-arrival studies of pulsars; use of interstellar scattering phenomena to probe the magnetospheres and orbits of radio pulsars; and the prospects for low frequency interferometry from space.

Finally, in Chapter Eight, the agendas from two previous meetings on interstellar scattering are presented for historical purposes. Neither of these meetings—held in 1974 and 1986—led to any published account. The antecedents of recent work may be seen in the 1974 agenda, yet it is clear that the theory and phenomena, and the relation of phenomena to other areas of astrophysics, are much more far reaching than anyone would have predicted.

J. M. Cordes, Ithaca
B. J. Rickett, La Jolla
D. C. Backer, Berkeley

I. SCINTILLATION PHENOMENA AND THEORY

INTRODUCTION TO THE OBSERVABLES IN INTERSTELLAR RADIOWAVE PROPAGATION

B.J.RICKETT
Dept of Electrical and Computer Engineering,
University of California, San Diego.

ABSTRACT

Radio waves travelling through the irregular plasma density in the interstellar medium are distorted in various ways. The differing observable quantities are introduced and the major conclusions that can be drawn from each about the wave-number spectrum of plasma density are discussed. Distinction is drawn between diffractive effects on scales 10^6-10^8m and refractive effects on scales of 10^{10}-10^{12}m.

INTRODUCTION

It is a pleasure to open this workshop on radiowave scattering in the interstellar medium. I would like particularly to welcome Professor Tony Hewish, both because he has travelled the furthest and because, as you all know, he was responsible for bringing pulsars into our field of view. I will introduce the observational side of the subject, addressing my remarks primarily to those who are not already experts.

The discovery of pulsars opened a major new probe of the interstellar medium. The most direct pulsar observables are the frequency-dependent arrival times and polarization angles. The two coefficients of the frequency dependence are, respectively, dispersion measure (DM) and rotation measure (RM). For propagation through the ionized interstellar plasma at radio frequen-cies well above the local plasma frequency and the local gyro frequency, these give the following estimates:

$$DM = \int^L N_e \, dz, \text{ (often given in units of cm}^{-3} \text{ pc) and } RM = \int^L N_e \, \mathbf{B}.\mathbf{dz}. \quad (1)$$

Evidently, the ratio RM/DM gives an average of the line of sight component of magnetic field, weighted by the electron density. Observation in the 21cm line of neutral hydrogen with the pulse on and off allows both emission and absorption spectra on the same line of sight to be determined; given a model of the rotation and distribution of HI, the identification of which rotation features are present in the absorption leads to constraints of the pulsar's distance L. See Weisberg at al. (1980) for examples of this type of measurement. Given the distance, the dispersion measure leads to a simple estimate of average electron density along the line of sight to each pulsar. A typical value is 0.03 electrons per cm^{-3} (see Lyne et al., 1985 ; Manchester and Taylor, 1981) for detailed models of the electron distribution in the ISM).

The less directly-interpretable quantities reflect the fact that the electron distribution is irregular at surprisingly small scales. This was not anticipated before the discovery of pulsars, but similar plasma irregularities were already known in the ionosphere and the interplanetary medium. The goal of my presentation is to review the various observable quantities that yield information on the irregularities in the interstellar plasma density; I would like to make the discussion accessible to those of you who are not afficionados of scattering theory; the intricate details will assuredly follow in later talks. Our hope for this meeting is that it will stimulate more interaction between radio observers and theorists concerned with the interstellar medium (ISM).

It is convenient to divide radio propagation phenomena into refractive and diffractive regimes, corresponding to effects due respectively to large (10^{12} - 10^{10} m) and small (10^8 - 10^6 m) scales in the medium. The dividing scale is roughly that of the radius of the Fresnel zone at a typical distance to the scattering irregularities.

DIFFRACTIVE SCATTERING

<u>Angular Scattering.</u> Spatial deviations in the refractive index cause phase modulations as the waves travel through the ISM. The resulting wavefront can be analysed into a spectrum of plane waves, the width of which we call the scattering angle (θ_s). A point or plane wave source thus suffers angular broadening. In radio astronomy angular structure is studied with an interferometer. If a pair of antennas are at vector positions s and $s+\sigma$, the interferometer output gives the amplitude and phase of the fringe visibility: $E(s) E^*(s+\sigma)$. If this complex quantity is averaged sufficiently, we find the ensemble average visibility function: $<E(s) E^*(s+\sigma)> = \exp[-0.5\, D_\phi(\sigma)]$, which is independent of the position coordinate s, if the medium is statistically stationary. For discussion of what constitutes sufficient averaging - a matter of some subtlety - see subsequent papers in these proceedings.

The quantity $D_\phi(\sigma)$ is the structure function of geometric phase - called the wave structure function in the optical literature - defined as follows:

$$D_\phi(\sigma) = < [\, \phi(s) - \phi(s+\sigma)\,]^2 > \quad ; \qquad (2)$$
$$\text{where } s = (x,y)\,,\quad \phi(x,y,L) = r_e\, \lambda \int^L N_e(x,y,z)\, dz\,,$$

with $N_e(x,y,z)$ as the deviation in electron density from its mean, λ is the (free-space) wavelength and r_e is the classical electron radius (2.82×10^{-15} m). Here ϕ is the phase deviation calculated on straight line path from the source to the observer, assuming that the observing frequency is everywhere much greater than the local plasma frequency.

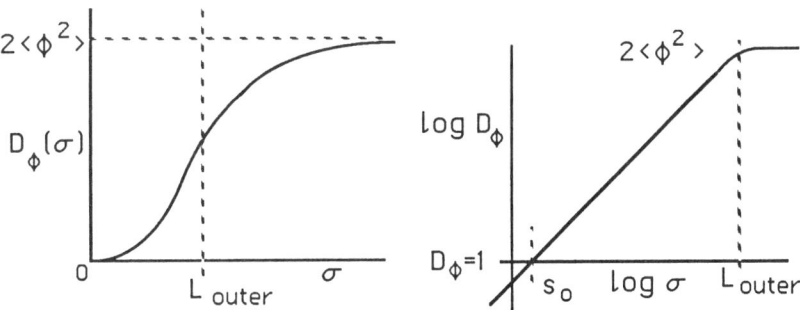

Figure 1 Structure Function of geometric phase on linear and log scales.

The structure function of ϕ is sketched schematically on linear and logarithmic scales in Figure 1. $D_\phi(\sigma)$ can be expressed in terms of the phase correlation function R_ϕ by $2<\phi^2> - 2 R_\phi(\sigma)$. It increases with σ from the origin, and at large displacements it saturates at a value equal to twice the phase variance ; we can define the outer scale by the value of σ at which it crosses half that value. On logarithmic scales we can examine the behavior on smaller scales; for strongly scattering media, we define a scale where $D_\phi(\sigma)$ equals unity. This is called the field coherence scale s_o which is the spatial separation across which an rms phase difference of 1 radian exists and at which the visibility function falls to $1/\sqrt{e}$.

Since the visibility function is a 2-D Fourier transform of the source brightness distribution, the uncertainty relation between widths in the two domains suggests that we define the width of the scattered image (scattering angle) by $\theta_s = (k\, s_o)^{-1}$. Of course, $D_\phi(\sigma)$ is determined by the statistical nature of the density irregularities. From its relation to the correlation function and the relation of phase to N_e, we obtain:

$$D_\phi(\sigma) = 4\pi\, \lambda^2\, r_e^2 \int^L \int^\infty [1-\cos(\kappa\cdot\sigma)]\, P_N(\kappa_x, \kappa_y, \kappa_z=0)\, d^2\kappa\, dz \qquad (3)$$

$P_N(\kappa)$ is the spectrum of the density deviations at 3-D wavenumber κ; it is this function that we wish to study - it provides the statistical description of the scales present in the density distribution. In the early studies of interstellar scintillation (e.g. Scheuer, 1968; Rickett, 1970) a gaussian form was assumed. The observations were used to estimate the single gaussian scale; it seemed a remarkable coincidence that the number obtained was close to the "Fresnel" scale $(L/k)^{0.5}$. The presence of a range of scales was proposed by Salpeter (1969), though he did not express it as a power law wavenumber spectrum. Lovelace (1970) proposed a power law form for the spectra of both the interstellar and interplanetary plasma density. Lee and Jokipii (1976) suggested a power law spectrum over many decades of wavenumber like that of neutral turbulence - and made the bold extrapolation from the "microscales" responsible for radio scattering all the way to the parsec scale of interstellar clouds. They found that if the spectrum was extrapolated down to a wavenumber $\sim(1\text{ pc})^{-1}$ the variance in the density is on the order of the square of its mean, and so suggested the cloud size as an outer scale of a "turbulent" process. These ideas were explored further by Armstrong, Cordes and Rickett (1981). There have been few attempts to examine this theoretically; Higdon (1986) has proposed the treatment of density structure as a passive scalar in a turbulent flow, likened to that in a neutral fluid. The possibility of magnetohydrodynamic turbulence in the ISM has been considered by Montgomery et al. (1987; also in these proceedings). Other theorists have found specific damping mechanisms at a variety of length scales, making it very hard to accept an energy cascade over such a wide range (see papers by Max and by Zweibel in these proceedings for other theoretical discussions).

A common assumption in analysis of interstellar scattering data has been that the spectrum can be factored into a distance dependence times a power law wavenumber form:

$$P_N(\kappa) = C_N^2(z) \kappa^{-\beta} \quad \text{for } \kappa_{outer} < \kappa < \kappa_{inner} \tag{4}$$

with saturation in P_N for wavenumbers below and a steep cut-off for wavenumbers above this range. For such spectra there is a range of scales for which $D_\phi(\sigma)$ varies as a power of σ as in Fig 1 and can be approximated as $(\sigma/s_0)^\alpha$, where $\alpha \equiv \beta - 2$. Note the Kolmogorov value $\beta = 11/3$ and $\alpha = 5/3$. For such spectra the visibility becomes:

$$< E(s) E^*(s+\sigma) > = \exp[-(\sigma/s_0)^\alpha], \tag{5}$$

where $(s_0)^{-\alpha} \propto \lambda^2$ gives the exact wavelength scaling. From the baseline dependence of VLBI visibilities, we should therefore be able to detemine α. Since the wavelength scaling is independent of α in the visibility domain, observations over several wavelengths can only be used to increase the range of baselines with detectable visibilities and not to detemine the value of α. The question of whether the averaging in a typical VLBI experiment is sufficient to approximate the ensemble average, is tricky. If the exponent $\alpha<2$ ($\beta<4$), the answer is probably yes, but for steeper exponents the answer is probably no. Attempts to distinguish between exponents above or below this limit are therefore subject to interpretational difficulties at present. This question is still the subject of debate (see the papers by Spangler, Goodman and Narayan, Spangler et al. and others in these proceedings). The scattering angle can be calculated and for ($0<\alpha<2$) its scaling with wavelength and distance (in a uniform scattering medium) is given by:

$$\theta_S = (k\, s_0)^{-1} \propto \lambda^{(1+2/\alpha)} L^{(1/\alpha)} \tag{6}$$

Typical values cover 10^{-4} to 1 arcsec, requiring long baselines. For $2<\alpha<4$ see the paper by Narayan in these proceedings for the scalings. Note, however, that the wavelength scaling of θ_S is best examined as visibility versus baseline as mentioned above. Cordes, Ananthakrishnan and Dennison (1984) showed that θ_S only increases erratically with decreasing galactic latitude. They concluded that the "strength" of the density irregularities is not uniformly distributed, but is highly clumped, particularly at the low galactic latitudes toward the inner galactic plane, where θ_S is large enough to be measurable.

<u>Temporal Broadening of Pulses.</u> The pulses from pulsars are broadened as they travel through the irregular ISM, since the waves that arrive at an angle θ_S have evidently travelled further (see Cronyn 1970). For plane waves on a single scattering screen at distance L the extra delay is $\tau_S \sim L\theta_S^2/2c$. If there is an extended medium with total distance L, then $\tau_S \sim b\, L\theta_S^2/2c$, where b depends on the distribution of scattering versus distance. For such extended scattering media and ($0<\alpha<2$), the scaling is:

$$\tau_S \propto \lambda^{(2+4/\alpha)} L^{(1+2/\alpha)} \tag{7}$$

Measurements of pulse broadening have been made for many pulsars at many wavelengths and distances. The scaling is close to λ^4; the detailed results will be discussed below with the frequency decorrelation measurements. Not much has been done with the shape of the scattered pulse, but it has the potential to help determine the distribution of scattering along the line of sight (see Williamson 1974).

Visibility Fluctuations. Estimates of the visibility that are on too short a time scale to approximate an ensemble average, could be used to help study the medium, but not much has been done with them at present. From the theoretical standpoint, the variability of the second moment requires understanding of the fourth moment of the field.

Diffractive Intensity Fluctuations. Intensity fluctuations are also described by fourth order correlations of the field. The first evidence for fine scale structure in the ISM came from variations of pulsar intensities, soon after their discovery. Twenty years ago, while studying pulsars for my Ph.D. thesis, I was struck by the increase in the percentage intensity variability as the observing bandwidth was reduced (to combat the smearing effects of interstellar dispersion). This was the first glimpse of a phenomenon (diffractive interstellar scintillation - DISS) that has been much studied since; it causes pulsar amplitudes to fluctuate over the order of minutes to hours, independently over frequency widths that are characteristic of a particular pulsar and frequency, typically in the range of 10KHz to 10 MHz. Fig 2 shows a dynamic spectrum plot (intensity versus frequency and time).

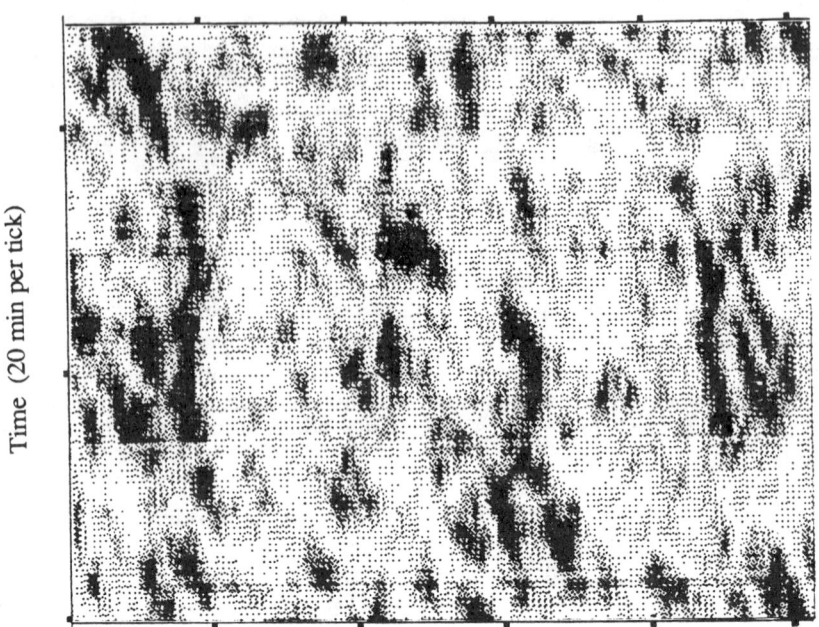

Figure 2. Intensity plot of a dynamic spectrum of diffractive scintillation of pulsar P0329+54 at 408 MHz. Horizontal scale is 0.5 MHz per tick; vertical 20 min per tick.

If a receiver bandwidth were to include several such independent "scintles", the percentage variation would clearly be reduced. We can interpret the temporal variations as spatial variations, mapped into time by the velocity of the observer ($\sim$ 50-250 Km/s) with respect to the diffraction pattern, providing that any temporal rearrangement of the pattern is typified by slower velocities.

The depth and time scale of the variations can be described by the spatial autocorrelation of intensity. In a theoretical description of DISS, we recognize that at scales near the field coherence scale in very "strong scattering", the field approximates a normal random variable and the intensity correlation is approximately the square of the second moment field of field given above. With this insight we see that the intensity pattern will have essentially 100% fluctuations and a spatial scale approximately s_0. Thus the time scale is

$$\tau_I \sim s_0/V \propto \lambda^{-2/\alpha} L^{-1/\alpha} \qquad (8)$$

The frequency scale is essentially the reciprocal of the temporal broadening time τ_s, ie

$$\Delta f_I \sim (2\pi\tau_s)^{-1} \propto \lambda^{-(2+4/\alpha)} L^{-(1+2/\alpha)} \qquad (9)$$

Again these are for $0 < \alpha < 2$; see Narayan's paper for larger values of α. Examples of observations of these quantities are shown in Figures 3 and 4, taken from the literature. The results can be summarized by noting that whereas the wavelength scalings are consistent with $\alpha \sim 1.8 \pm .15$ (comparable with the Kolmogorov value $\alpha=5/3$), the distance scalings are more variable and are steeper than expected for an extended medium. The accepted interpretation of this discrepancy is that the medium is not uniform. See Rickett (1977) and especially Cordes, Weisberg and Boriakoff (1985) for reviews of the observations. They interpreted the data as a power law spectrum with an exponent near the Kolmogorov value in a uniform disc $\pm$ 500pc thick, on which is superimposed a highly clumped distribution of very strongly scattering regions, concentrated within $\pm$ 100 pc of the inner galactic plane. The mean free path for a line of sight to encounter such enhanced scattering is about 1 Kpc; the local variance in the microstructure must be greater by factors of up to 10^5 to explain the increases of up to 10^4 in the integrated scattering strength.

Several questions posed by the diffractive scintillation results remain to be answered and are discussed in several papers in these proceedings.

- What is the physical origin of the density irregularities?
- Which of the ionised phases of the ISM is responsible?
- Is the same phase also responsible for the pulsar dispersion?
- Are the density irregularities related to magnetic irregularities involved in cosmic ray diffusion?
- Is there a turbulent cascade from large to small scales? or from small to large?
- If it is turbulence is it "active", or "passive" or "fossil" turbulence?
- What are the regions of the inner Galaxy that cause such enhanced turbulence?
- Also note the presence of a similar power law spectrum in the density of the solar wind, and also in the magnetic field spectrum. It seems that this is a common state for an astrophysical plasma, and its theoretical explanation warrants our attention.

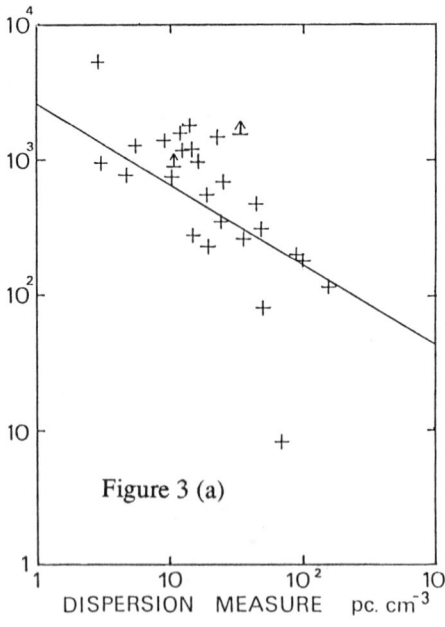

Figure 3 (a)

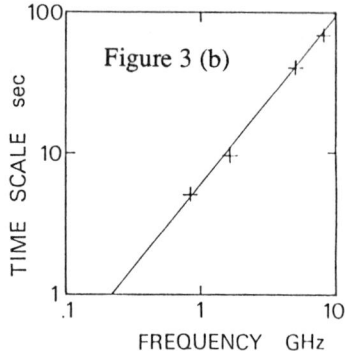

Figure 3. Measurements of τ_I: (a) at 400 MHz against pulsar dispersion measure and (b) for P0833-45 against frequency. (From Backer, 1975)

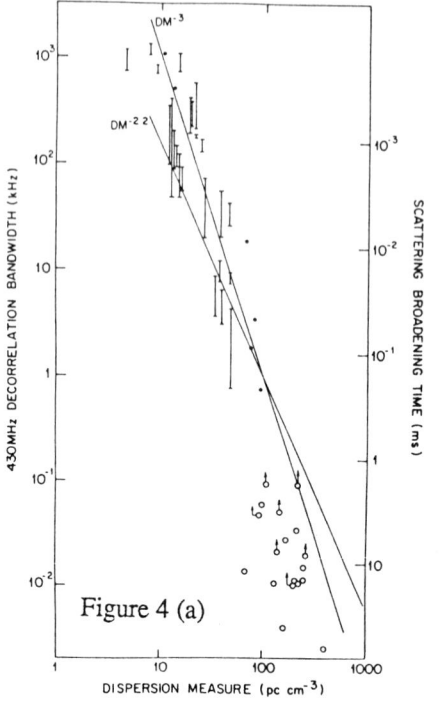

Figure 4 (a)

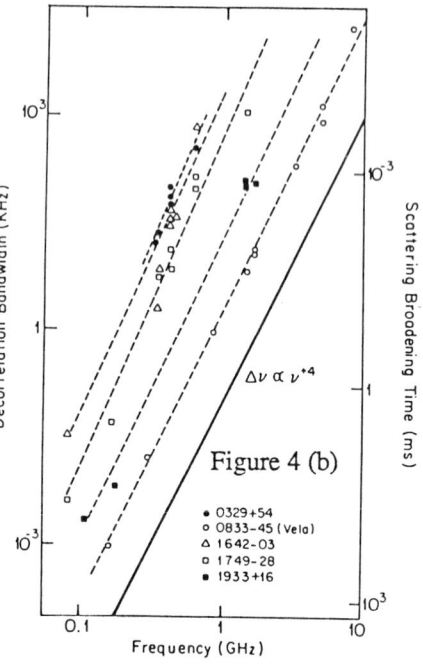

Figure 4. Measurements of Δf_I: (a) at 430 MHz against pulsar dispersion measure and (b) for 5 pulsars against frequency. (From Cordes et al. 1985)

REFRACTIVE INTERSTELLAR SCINTILLATION (RISS)

<u>Slow pulsar intensity fluctuations</u>. Pulsars vary in intensity by 10-100% rms over time scale τ_R, which is of the order of days to months. Sieber (1982) noticed that τ_R tends to increase with distance and with wavelength. This can be interpreted as the "refractive branch" of strong interstellar scintillation, a regime that had been noted theoretically (Prokhorov et al. 1975) and observed optically (Coles and Frehlich, 1982). This is now referred to as refractive interstellar scintillations (RISS).

The theory can be stated simply in terms of the "scattering disc". This is the lateral extent $L\theta_S$ of the region contributing to the signal received at a single point. It controls the largest scale irregularity that can influence the intensity. With a power law spectrum of density there is an increasing amplitude of irregularity with increasing scale, and so $L\theta_S$ is the largest scale present in the intensity pattern.

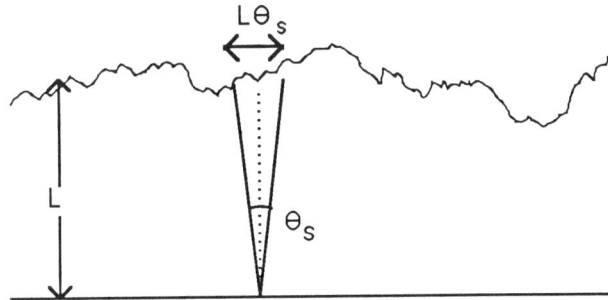

Figure 5. Sketch of scattering disc - the extent of region at screen that influences field at distance L.

We can think of the resulting intensity modulations as refractive in nature, caused by partial focusing and defocusing due to curvature of the wavefront averaged over the scattering disc. The sketch in Figure 6 suggests the direction of refraction from the large scale phase gradients present in the medium. The displacement of the diffractive pattern due to refraction is $L\theta_{Rx} = (\partial\phi/\partial x)(L/k)$ is indicated below at two frequencies ($f_1 < f_2$). The frequency dependence of the refraction is discussed below. The diffractive intensity pattern is also sketched as a series of narrow peaks in the intensity - displaced laterally by $L\theta_{Rx}$. The diffractive peaks are thus crowded together at positions of partial refractive focussing and spread apart at positions of defocussing. When averaged over scales larger than the diffractive scale, the resulting refractive intensity variation is I_R with rms modulation index of m_R, which is perhaps 5-50% for the Kolmogorov spectrum. The temporal variations would be given by the mapping at velocity V of the observer, with respect to the intensity pattern. Of course, the two dimensional nature of the scattered wave front complicates the mapping, since the velocity will not in general be parallel to the local refractive displacement. Nevertheless, typical refractive fluctuation times can be estimated as $\tau_R \sim L\theta_S/V$.

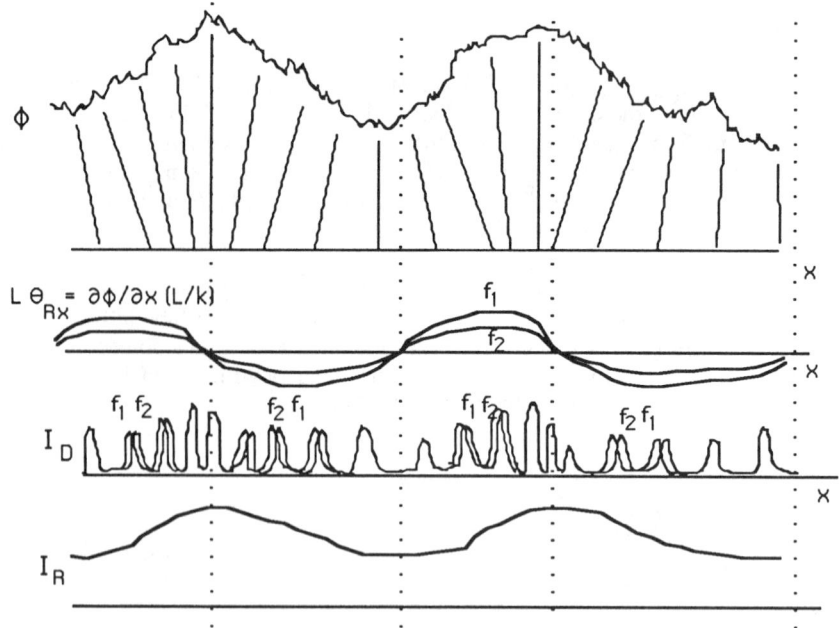

Figure 6. Schematic relations between the refractive and difffractive intensities.

Putting in the estimated expression for θ_s with distance and wavelength:

$$\tau_R \sim 17 \text{ days } (L_{Kpc} \theta_{mas} / V_{100Kms}) \propto \lambda^{(1+2/\alpha)} L^{(1+1/\alpha)}$$

(10)

The scalings with distance and wavelength agree approximately with the data on slow pulsar variability compiled by Sieber (1982) as discussed by Rickett, Coles and Bourgois (1984) and shown in Figure 7. The basic ideas of refractive intensity scintillations seem to be verified. However, the percentage modulation (m_R) at the refractive time scale appears to be larger than that predicted for the Kolmogorov spectrum model. This result lead Blandford and Narayan (1985) to propose a spectrum model with an exponent greater than 4 (see also Goodman and Narayan, 1985 and Romani et al. 1986), for which m_R is about unity. Figure 8 illustrates three possible spectra that have been advanced to explain the observed values of m_R. Coles et al. (1987) analyzed the influence of an inner scale approximately at 10^9 m, cutting off at wavenumbers above the reciprocal of that scale. Another possibility is a Kolmogorov spectrum with an enhancement in spectral density below a particular wavenumber; Cordes, Pidwerbetsky and Lovelace (1986). What all three of these spectra have in common is a greater ratio between the spectral densities at the refractive and diffractive wavenumbers than exists for the Kolmogorov case. There remains, however, the need for better measurements of m_R, such as the recent observations by Stinebring and Condon (1988).

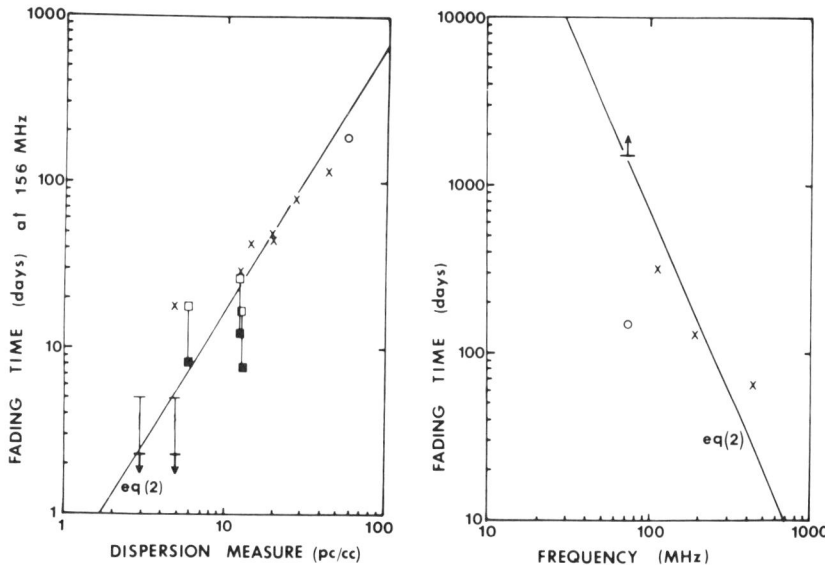

Figure 7. Measurements of τ_R : (a) at 156 MHz against pulsar dispersion measure and (b) for P0531+21 against frequency. (From Rickett et al. 1984)

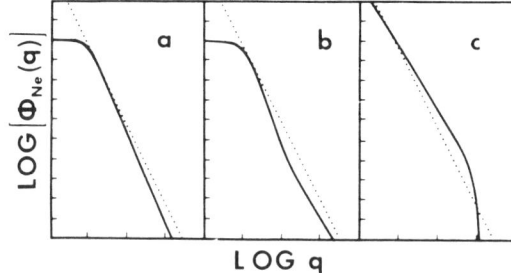

Figure 8. Possible models for the spatial spectrum of electron density. (a) The steep spectrum with $\alpha > 2$; (b) Kolmogorov spectrum with low wavenumber enhancement; (c) Kolmogorov spectrum with inner scale cut-off. (from Coles et al. 1987).

RISS for radio sources - other than pulsars. Only pulsars have small enough angular extent to show diffractive intensity fluctuations; however, the cut-off for RISS is much less stringent. Thus galactic and extra-galactic sources can exhibit RISS if they have very small angular diameters. This appears to be the explanation for the intensity variations of "meter wavelength variables", centimeter wavelength "flickering" among others (Rickett, 1986). It is not clear whether all variations of these sources are RISS, but the phenomenon is a major factor. The possibility of mixed intrinsic and scintillation variations makes interpretation difficult and no new constraints on the interstellar irregularities have yet come from such observations.

Position wandering. The variable angle of refraction should cause the position of a radio source to appear to be displaced from its true position. The rms amplitude of the position fluctuation and the time scale for it to fluctuate depend critically on the density spectrum, being larger and slower for spectral exponents steeper than 4. Potentially this is a useful diagnostic for such steep spectra. Few measurments exist at present, though interesting measurements by Gwinn et al. are reported in these proceedings.

Refractive effects in pulsar dynamic spectra. Since the earliest dynamic spectra of diffractive scintillation from pulsars, sloping bands in the frequency-time plane have been noticed. These can be understood as refractive steering of the diffractive intensity pattern. As indicated in Figure 6, the refractive displacement of the pattern is frequency dependent (varying as frequency $^{-2}$). Hence the diffractive intensity peaks will be displaced between frequencies $f_1 < f_2$. If the x-direction is parallel to the observer's motion (figure 6), a spectrum will show sloping features, in frequency-time, changing sign as the refractive x-displacement changes sign. Various authors have investigated these phenomena; see Shishov (1974), Roberts and Ables (1982), Hewish et al. (1985), Cordes and Wolszczan (1986). A further effect is that the apparent bandwidth of diffractive scintillation is reduced by the presence of refractive slopes and so fluctuates over time scales of τ_R.

In addition to the common occurrence of sloping bands, observers have found frequent occasions of overlapping slopes with opposite sign and also occasions of periodic parallel bands. Hewish (1980) was the first to argue that these were inconsistent with a Kolmogorov spectrum, a view supported by Cordes and Wolszczan (1986), who found intervals of remarkably periodic "fringe patterns" crossing the dynamic spectra. Both phenomena can be understood as pairs of refracted beams (rays) that intersect and may interfere with each other. It remains to determine whether such behavior is caused by the same distribution of large scale irregularities as cause the RISS (for example due to behavior near a partial focus) or whether the highly periodic bands are rarer occurrences like the extreme scattering events, discussed below. What is needed is a statistical estimator for these phenomena and a study of their rate of occurrence versus distance, wavelength and direction through the galaxy.

Pulsar arrival time fluctuations. Just as the total electron content between us and a pulsar determines the total dispersion delay, the variation in electron content can cause a frequency dependent fluctuation in delay (seen as a slowly varying arrival time). Lovelace (1970) and Rickett (1977) gave the theory to first order, noting that the timing fluctuation in a data span of T (several years typically) is determined by density scales on the order of VT. More specifically $\tau_g = \partial\phi/\partial\omega = -\phi/\omega$ and the rms delay variation in T would be given by:

$$\tau_g(T)^2 \sim D_\tau(T) = D_\phi(VT)/(\omega^2) \tag{11}$$

See Blandford et al. (1984) for a detailed discussion of the various possible propagation influences on pulsar timing data. Very few timing observations have had the dual frequency coverage needed to identify dispersive delay variations. Rawley (1986) analyzed timing observations of the "millisecond" pulsar (1937+21) made at 1.4 and 2.5 GHz. Figure 9 shows his estimates of dispersion variation over 650 days.

The structure function of dispersion was estimated from these data and converted into an estimate of $D_\phi(s)$ which is shown in figure 10. The diagram shows these estimates of D_ϕ together with an estimate of where $D_\phi(s_0) = 1$, based on a measurement of the diffractive scintillation bandwidth reported to be 1.3 MHz at 1420 MHz by Rawley. The relation between Δf_I and s_0 was used, appropriate for a screen midway between the pulsar and the earth at a total distance of 4 kpc. Departures from the midpoint would increase the value inferred for s_0 as indicated by the bar extending to the right. Lines of slope 2 and 5/3 are shown; The conclusion here is that the data are only compatible with α less than about 1.8, ie excluding the steep power law spectral models. If there were an inner scale at 10^9 m the slope would be 2 below that scale and become 5/3 above it. This line is close to the data but lies slightly above. The result presented is interesting because of the four orders of magnitude difference in scale between the two estimates, and the fact that the technique is straightforward to interpret. It should be remembered, however, that the dispersion variations come from small timing differences up to a maximum of 2 μs; such extremely small values are very difficult to measure.

Interstellar Focusing Events. A new ingredient has been added to the phenomenology of interstellar scattering, with the discovery by Fiedler et al. (1987) of remarkable time profiles of intensity in the NRL source monitoring observations at 3 and 8 GHz. The profiles last tens of days during which time a source appears alternately enhanced and diminished in intensity by as much as 50% in a sequence of apparently deterministic rather than random variations. (See the paper by Fiedler et al. in these proceedings). The suggested explanation is the interposition of large discrete ionized structures along the line of sight. Two mechanisms have been advanced; scattering (Fiedler et al. 1987) or partial focusing (Romani et al.1988). The structures might be nearer spherical or planar, with an alignment along the line of sight occurring rarely. The questions raised by these events ("extreme scattering events" or "interstellar focussing events") add new intriguing questions to our studies of the ISM. Romani et al. (1988) suggest that they are related to the structures that cause the (rare) intervals of highly periodic fringes in pulsar dynamic spectra. Are they perhaps part of the random wavenumber spectrum of

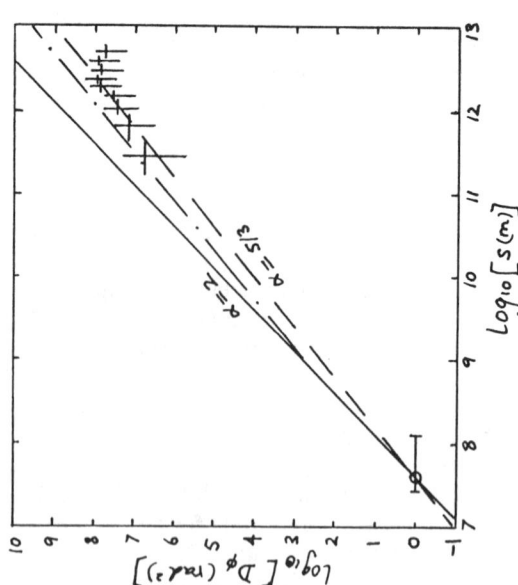

Figure 10. Estimates of the structure function of phase derived from the data of Figure 9. The point at $D_\phi = 1$ comes from a decorrelation bandwidth measured at 21cm (also by Rawley, 1986). Notice that the line of slope 2 (corresponding to $\alpha \geq 2$) lies a factor of 20 above the observations at a scale of 10^{12} m. The dot-dash line is for a model with an inner scale; it breaks at 10^9 m from a slope of 2 to 5/3 (the Kolmogorov value).

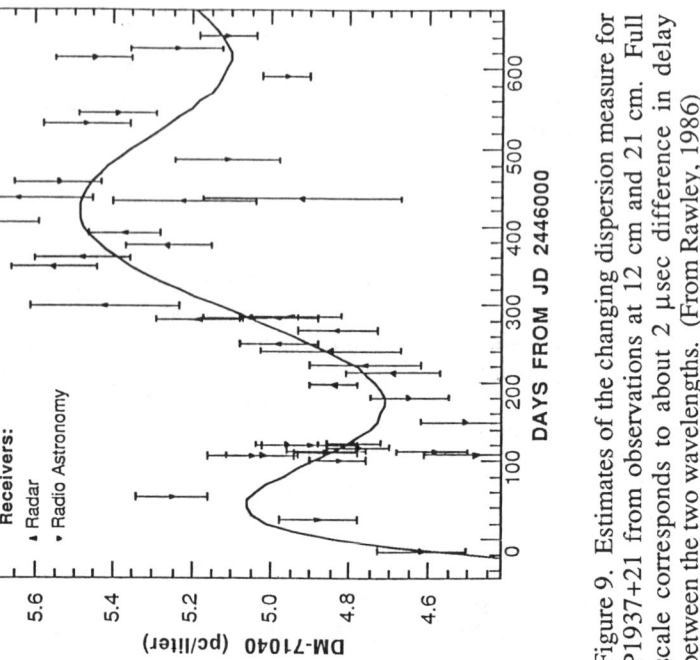

Figure 9. Estimates of the changing dispersion measure for P1937+21 from observations at 12 cm and 21 cm. Full scale corresponds to about 2 µsec difference in delay between the two wavelengths. (From Rawley, 1986)

electron density, discussed earlier? Or are they additional discrete structures that occasionally intervene? If they are indeed describable as part of the wavenumber spectrum, it seems possible that we are dealing with a Kolmogorov spectrum, with an enhancement at the low wavenumbers rather than a cut-off at the high wavenumbers.

QUESTIONS AND CONCLUSIONS

The study of radio propagation in the ionized ISM has matured from a backwater topic of interest to a few pulsar observers to one of interest to a broader community of radio astronomers concerned with the entire spectrum of variable sources and also to those concerned with the ultimate VLBI resolution. A variety of observable phenomena has lead to a steady advance in our understanding of the interstellar electron density spectrum and has demonstrated the presence of a power law density spectrum over as much as five orders of magnitude, with an exponent close to that of Kolmogorov turbulence in a neutral fluid. (See the papers by Narayan and by Spangler for detailed discussions of how these conclusions are reached). This has raised the issue of whether there is a turbulent cascade of energy from the large scales to dissipation at the small scales? What is the physics of such plasma turbulence? Are the magnetic field and velocity spectra of the same form? The spatial distribution of the density irregularities is also of interest; against the background distribution in a fairly wide ($\pm$ 500pc) disc, there are localized enhancements in the level of the density spectrum by as much as five orders of magnitude in the inner galactic plane. There is no consensus on the physical site of such enhancements but this question as well as many others were the subject of much discussion at the workshop.

I thank J.Cordes, L.Rawley, D.Backer for permission to reproduce their data. I thank Bill Coles for computing the structure function in Figure 10 and for valuable discussions.

REFERENCES

J.W.Armstrong, J.M.Cordes and B.J.Rickett, Nature, **291**, 561, (1981).
D.C.Backer, Astrophys. J. **43**, 395, (1975).
R.D.Blandford and R.Narayan, Mon. Not. Roy. Astro. Soc. **213**, 591, (1985).
R.D.Blandford and R.Narayan and R.W.Romani, J.Astrophys. Astron. **5**, 369, (1984).
W.A.Coles and R.G.Frehlich, J. Opt. Soc. Am. **72**, 1042, (1982).
W.A.Coles, R.G.Frehlich, B.J.Rickett and J.L.Codona, Astrophys. J. **315**, 666, (1987).
J.M.Cordes, S.Ananthakrishnan and B.Dennison, Nature, **309**, 689, (1984).
J.M.Cordes, J.M.Weisberg and V.Boriakoff, Astrophysical J. **288**, 221, (1985).
J.M.Cordes and A.Wolszczan, Astrophys. J. Lett. **307**, L27, (1986).
J.M.Cordes, A.Pidwerbetsky and R.M.Lovelace, Astrophys. J. **310**, 737, (1986).
W.M.Cronyn, Science, **168**, 1453, (1970).
R.L.Fiedler B.Dennison, K.J.Johnston and A.Hewish, Nature, **326**, 675, (1987).
J.Goodman and R.Narayan, Mon. Not. Roy. Astro. Soc. **214**, 519, (1985).
A. Hewish, Mon. Not. Roy. Astro. Soc. **192**, 779, (1980).
A. Hewish, A. Wolszczan and D.A. Graham, Mon. Not. Roy. Astro. Soc. **213**, 167, (1985).

J.C.Higdon, Astrophys. J. **309**, 342, (1986).
L.C.Lee and J.R.Jokipii, Astrophys. J. **206**, 735, (1976)
R.V.E.Lovelace, Ph.D. Thesis, Cornell University, (1970).
A.G.Lyne, R.N.Manchester and J.H.Taylor, Mon. Not. Roy. Astro. Soc. **213**, 613, (1985)
R.N.Manchester and J.H.Taylor, Astron. J. **86**, 1953, (1981).
D.Montgomery, M.R. Brown and W.H.Matthaeus, J. Geophys. Res. **92**, 282, (1987).
A.M.Prokhorov, F.V.Bunkin, K.S.Gochelashvily and V.I.Shishov, Proc. I.E.E.E. **63**, 790, (1975).
L.Rawley, Ph.D. Thesis, Princeton University, (1986).
B.J.Rickett, Mon. Not. Roy. Astro. Soc. **150**, 67, (1970).
B.J.Rickett, Ann. Rev. Astron. Astrophys. **15**, 479, (1977).
B.J.Rickett, Astrophys. J. **307**, 564, (1986).
B.J.Rickett, W.A.Coles and G.Bourgois, Astron. and Astrophys. **134**, 390, (1984).
J.A.Roberts and J.G.Ables, Mon. Not. Roy. Astro. Soc. **201**, 1119, (1982).
R.W.Romani, R.Narayan and R.Blandford, Mon. Not. Roy. Astro. Soc. **220**, 19, (1986).
R.W.Romani, R.D.Blandford and J.M.Cordes, Nature, **328**, 324, (1988).
E.E.Salpeter, Nature, **221**, 31, (1969).
P.A.G.Scheuer, Nature, **218**, 920, (1968).
W.Sieber, Astron. and Astrophys. **113**, 311, (1982).
V.I.Shishov, Soviet Astron. **17**, 598, (1974),
D.Stinebring and J.J.Condon, Astrophys. J. preprint, (1988).
I.P.Williamson, Mon. Not. Roy. Astro. Soc. **166**, 499, (1974).
J.M.Weisberg, J.M.Rankin and V.Boriakoff, Astron. and Astrophys. **88**, 84, (1980).

FROM SCINTILLATION OBSERVATIONS TO A MODEL OF THE ISM—THE INVERSE PROBLEM

Ramesh Narayan

Steward Observatory
University of Arizona
Tucson, AZ 85721

ABSTRACT

The inverse problem consists of two parts: (i) To establish the *form* of the electron density fluctuation spectrum. (ii) To determine the *strength* of fluctuations in different regions of the ISM.

(i) Based on a variety of observations of scintillation-related phenomena in radio pulsars and compact galactic and extragalactic radio sources, it appears that the density fluctuation spectrum for lengthscales shorter than $\sim 10^{14}$ cm is close to that predicted by Kolmogorov turbulence theory. Refractive scintillation is found to be unusually strong in several sources, suggesting either that there is enhanced fluctuation power at $\sim 10^{13}$–10^{14} cm or that the inner scale is larger than thought earlier, possibly as large as $\sim 10^{11}$–10^{12} cm.

(ii) Electron density fluctuations in the Galaxy consist of a strong clumpy component with a scale height ≤ 100 pc and a weaker nearly homogeneous component with a scale height ≥ 500 pc. The former component is particularly strong in the regions of the Galaxy interior to the sun and may be associated with HII regions and supernova remnants.

1. INTRODUCTION

Ever since the early discovery of radio pulsar scintillation and its explanation by Scheuer (1968) as an effect due to the interstellar medium (ISM), a variety of scintillation-related phenomena have been identified and studied, mainly with a view to understanding the ISM. In addition to radio pulsars, other compact galactic and extragalactic sources have also proved useful in this endeavor.

Much of the theoretical work in interstellar scintillation (ISS) has concentrated on what may be called the *forward problem*. This involves starting with a model of the electron density fluctuations in the ISM and predicting the magnitudes and the characteristics of various observables (e.g., Lovelace 1970, Gochelashvily and Shishov 1975, Rickett 1977, Goodman and Narayan 1985, hereafter GN, Romani, Narayan and Blandford 1986, Cordes, Pidwerbetsky and Lovelace 1986). The theory is well-developed for a power-law form of the density fluctuation spectrum, particularly when the scattering occurs in a narrow layer between the source and the observer—the "thin screen" case.

The *inverse problem*, in contrast, starts with the observations and seeks a density fluctuation model of the ISM. Unfortunately, there is no unique procedure to do the inversion. This is because ISS is dominated by scattering on two widely separated lengthscales, the diffractive scale $\sim 10^8$–10^9 cm and the refractive scale $\sim 10^{13}$–10^{14} cm, and is insensitive to fluctuations on other scales. Consequently, the problem of inverting the observations to estimate the entire power spectrum is ill-conditioned. In addition, the dynamic range of scales involved, which can exceed 10^4, makes direct inversion somewhat impractical. The usual approach to the inverse problem is instead through what may be termed the "forward-inverse" method where one solves the forward problem for a variety of models and seeks the best fit with the observations.

The inverse problem is traditionally divided into two parts:

(i) To determine the *form* of the density fluctuation spectrum as a function of lengthscale (or spatial wave-vector).

(ii) To estimate the *strength* of the fluctuations, i.e. the normalization of the spectrum, as a function of position in the ISM.

It is usual to assume that the two parts are independent of each other; we follow this practice by discussing (i) in section 2 and (ii) in section 3. However, it is possible, and even likely, that the form of the spectrum differs in different regions of the ISM.

2. FORM OF DENSITY FLUCTUATION SPECTRUM

In his seminal paper on ISS, Scheuer (1968) assumed that the density fluctuations in the ISM consist of blobs of a characteristic scale. The twin requirements of (i) strong scattering and (ii) multipath propagation constrained this scale to lie in the range 10^9–10^{13} cm. Scheuer's theory explained the origin of saturated intensity modulation in radio pulsars and also predicted dependences of scintillation observables with wavelength λ that were in approximate agreement with observations. However, small but significant discrepancies are now being found, and the single-scale model is no longer in use.

Lee and Jokipii (1975) took the important step of postulating a turbulent origin for the density fluctuations in the ISM. Invoking a Kolmogorov cascade, they suggested that the three-dimensional spectrum of electron density fluctuations takes the form of an extended power-law,

$$P_{3N}(q) = C_N^2 q^{-\beta}, \qquad \beta = 11/3, \tag{1}$$

where q is a spatial wave-vector and the normalization constant, C_N^2, measures the strength of the density fluctuations (sec. 3); the choice of the power-law exponent, $\beta = 11/3$, corresponds to the Kolmogorov scaling. Electron density fluctuations in the medium introduce transverse phase fluctuations in the wavefront from a distant radio source and it is easy to show that the two-dimensional transverse phase power spectrum has the same form as the three-dimensional density spectrum, i.e. eq. (1) is equivalent to $P_{2\phi}(q) \propto q^{-\beta}$. (In the thin-screen model, one needs to consider only phase fluctuations on the scattering screen, but for an extended medium one should keep track of both phase and amplitude variations as the wave propagates through the medium.) Lovelace (1970) made one of the earliest studies of scintillation with power-law phase spectra.

Quite apart from arguments based on turbulence theory, there are several independent *observational* reasons why a power-law spectrum with a range of scales is preferable to a single-scale model:

(i) Cordes, Weisberg and Boriakoff (1985, hereafter CWB) measured the decorrelation bandwidths, $\Delta\nu_{dc}$, of several pulsars as a function of λ. Fitting the dependences in the form

$$\Delta\nu_{dc} \propto \lambda^{-x}, \qquad (2)$$

they obtained a mean x of 4.45 ± 0.15 (error estimate from Cordes, private communication). Scheuer's single-scale spectrum predicts $x = 4$, while the spectrum in (1) predicts $x = 4.4$. Unfortunately, there does not exist a unique mapping from x to β. GN showed that $x(\beta)$ has a turning point at $\beta = 4$; therefore, $x = 4.4$ is consistent with both $\beta = 11/3$ and $\beta = 4.2$. Despite this complication, one can still say that the observed deviation of x from 4 rules out a single-scale model.

(ii) Rickett, Coles and Bourgois (1984) showed that slow intensity fluctuations on timescales of days to years in pulsars and compact extragalactic sources could be identified with refractive scintillation. This identification is now quite solid (Rickett 1986, Blandford, Narayan and Romani 1986) and the fact that strong refractive scintillation is observed in a variety of sources argues for widespread density fluctuations in the ISM on scales $\sim 10^{13}$–10^{14} cm. These scales differ by many orders from the scales $\sim 10^8$–10^9 cm that are thought to produce the usual diffractive scintillation, thus supporting the idea of an extended spectrum.

(iii) Armstrong, Cordes and Rickett (1981), following the initial suggestion of Lee and Jokipii (1976), demonstrated that the density fluctuation power on scales $< 10^{10}$ cm implied by ISS, and the power on scales of parsecs and kiloparsecs measured by direct observations, are both consistent with a single power-law of the form (1) with β not very different from $11/3$. This lends additional support to the concept of an extended turbulent cascade.

A scale-free power spectrum of the form (1) is clearly an idealization. In reality, the power-law must be terminated by cut-offs at an "inner scale", r_{in}, and an "outer scale", r_{out}. These scales may be neglected only when (i) r_{in}

is smaller than the smallest scale that influences the scintillation, which is the diffractive scale, r_{diff}, and (ii) r_{out} is larger than the largest scale sampled, which is either the refractive scale, r_{ref}, or vt_{obs}, where v is the velocity of the observer past the scintillation pattern and t_{obs} is the duration of the observations. These two conditions are not necessarily satisfied in ISS. Firstly, a turbulent cascade in the ISM could be arrested by several damping mechanisms on lengthscales much larger than r_{diff} (e.g. Cesarsky 1980), thus violating condition (i). Secondly, there is no proof that the turbulent energy input into the ISM really does take place on the parsec/kiloparsec scale that Lee and Jokipii (1976) and Armstrong et al. (1981) proposed (but see Jokipii, this volume, for a new argument in support of this hypothesis). It is conceivable that r_{out} is comparable to $r_{\text{ref}} \sim$ 1 AU, a natural scale for some processes in the ISM, in which case condition (ii) would be violated. Allowing for these possibilities, it is useful to consider the following general power-law model for the spectrum of transverse phase fluctuations,

$$P_{2\phi}(q) \propto (q^2 + q_{\min}^2)^{-\beta/2} \exp(-q/q_{\max}), \tag{3}$$

$$q_{\max} \sim 2\pi/r_{\text{in}}, \qquad q_{\min} \sim 2\pi/r_{\text{out}}. \tag{4}$$

Indeed, in order to be as general as possible within the power-law framework, we also allow the index β to be arbitrary. There are two reasons for this. Firstly, it is not clear that the Kolmogorov scaling, which was originally derived for incompressible fluids, is valid also for compressible MHD plasmas; although the work of Fyfe, Joyce and Montgomery (1977) and Higdon (1984) has allayed some of the concerns, this question still remains an issue. Secondly, in tackling the inverse problem of ISS, it is probably best to make the fewest theoretical assumptions regarding the ISM and to let the observations speak for themselves. This is the spirit in which the present review is written.

The form of the spectrum in eq. (3) assumes that the spectrum is isotropic. Higdon (1984) argues that dynamically important magnetic fields would cause density fluctuations in the ISM to be strongly anisotropic. The effects of such anisotropy on the characterstics of scintillation are just beginning to be analyzed (e.g. discussions by Narayan and Hubbard 1988, Narayan and Goodman 1988). Since at the present time there is no firm evidence for anisotropic ISS the rest of this review will assume isotropy.

We now define wave-vectors corresponding to the diffractive and refractive scale, $q_{\text{diff}} \sim 2\pi/r_{\text{diff}}$, $q_{\text{ref}} \sim 2\pi/r_{\text{ref}}$; note that $q_{\text{diff}} \gg q_{\text{ref}}$ for ISS. As mentioned earlier, when $q_{\max} > q_{\text{diff}}$ and $q_{\min} < q_{\text{ref}}$, the inner and outer scales play no role, and the characteristics of the scintillation are determined solely by the index β. Following Lovelace (1970) we will call such spectra as Type A when $\beta < 4$ and as Type B when $\beta > 4$. An alternative nomenclature that is often used is "shallow" for spectra with $\beta < 4$ and "steep" for $\beta > 4$.

When $q_{\max} < q_{\text{diff}}$ and $q_{\min} < q_{\text{ref}}$, density fluctuations on the diffractive scale are absent and some of the scintillation characteristics are modified. In an extension of Lovelace's classification, we will call these Type A1 and Type B1 spectra. Rickett et al. (1984) and Coles et al. (1987) have suggested

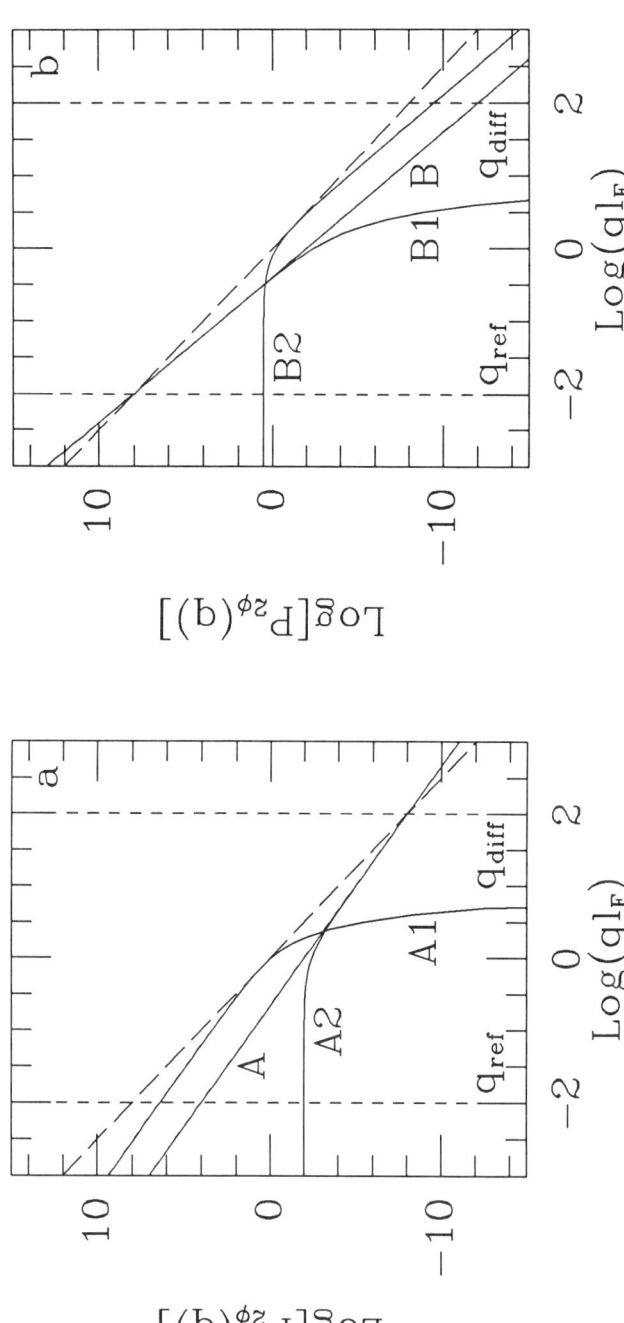

Figure 1. (a) Power spectra, $P_{2\phi}(q)$, of Types A, A1 and A2 as functions of reduced wave-vector ql_F, where l_F is the Fresnel length (GN). The sloping long-dashed line has a negative slope of 4 and the three spectra are less steep than -4 ("shallow" spectra). The vertical dashed lines indicate the refractive and diffractive scales. The scattering angles for Type A and Type A2 spectra are determined by the power $P_{2\phi}(q_{\text{diff}})$ at the diffractive scale and for a Type A1 spectrum by the power at the inner scale where the spectrum cuts off. The three spectra have been normalized to give the same scattering angle. (b) Spectra of Types B, B1 and B2. The spectra are more steep than slope -4 ("steep" spectra). The scattering angles for Type B and Type B1 spectra are determined by $P_{2\phi}(q_{\text{ref}})$ and for a Type B2 spectrum by the power at the outer scale. The three spectra have been normalized to give the same scattering angle.

that the ISM may be described by a Type A1 spectrum. For completeness, we define spectra with $q_{max} > q_{diff}$, $q_{min} > q_{ref}$ to be Type A2 and Type B2. As mentioned earlier, all these Types could refer to either the three-dimensional density spectrum or the two-dimensional phase spectrum, and we will use both interchangeably in our discussion.

Figure 1 illustrates the above definitions of the various types of spectra. We discuss these spectra in some detail in the succeeding subsections, describing their chief characteristics and reviewing their agreement with the observations.

2.1 Type A Spectrum (Shallow Spectrum)

The solid line in Fig. 2a shows the spatial spectrum of intensity fluctuations corresponding to a Type A spectrum with $\beta = 11/3$. The dominant peak at wave-vector q_{diff} is due to diffractive scintillation. In the strong scintillation regime, the diffractive intensity scintillations are "saturated", i.e. the area under the diffractive peak is unity, independent of the index β. Consequently, the amplitude of diffractive ISS does not constrain β in any way. However, the scalings with λ of several other diffraction-related observables do depend on β. For instance, the scintillation timescale, t_{scint}, the decorrelation bandwidth, $\Delta\nu_{dc}$, and the angular scatter-broadening, θ_{scatt}, vary as

$$t_{scint} \propto \lambda^{-2/(\beta-2)}, \quad \Delta\nu_{dc} \propto \lambda^{-2\beta/(\beta-2)}, \quad \theta_{scatt} \propto \lambda^{\beta/(\beta-2)}. \quad (5)$$

Early attempts to measure these scalings were not very accurate and merely suggested that β is of order 4. (Note that the scalings for $\beta = 4$ are identical to those predicted by the Scheuer spectrum.) However, recent work by CWB on several pulsars appears to rule out $\beta = 4$. As mentioned in the discussion following eq. (2), the data obtained by CWB are consistent with a Kolmogorov index, $\beta = 11/3$. Exceptions include the Vela pulsar, which displays $\Delta\nu_{dc} \propto \lambda^{-4}$ (Backer 1974), and the compact nonthermal source at the galactic center which scales as $\theta_{scatt} \propto \lambda^{2.0\pm0.1}$ (Lo et al. 1985); these scalings correspond to $\beta = 4$ (the latter actually being consistent with $3.8 < \beta < 4.1$).

The smaller peak at q_{ref} in Fig. 2a is due to refractive scintillation. The modulation index of refractive flux variations is given by (e.g. GN)

$$m_{ref} \equiv \frac{(\Delta F)_{ref}}{\overline{F}} = f(\beta) \left(\frac{q_{ref}}{q_{diff}}\right)^{(4-\beta)/2}, \quad (6)$$

where $f(\beta)$ is a constant of order unity. At meter wavelengths and for standard parameters, eq. (6) gives $m_{ref} \sim 0.1$–0.2 for $\beta = 11/3$. Many pulsars for which m_{ref} has been estimated seem to have intensity fluctuations with m_{ref} well in excess of 0.2 (see GN and references therein). This is probably the strongest argument against a Type A spectrum with $\beta = 11/3$, though it must be mentioned that the quality of some of the data is not very high. A Type A

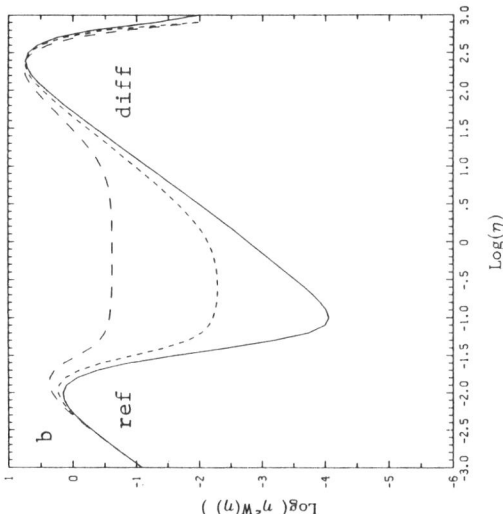
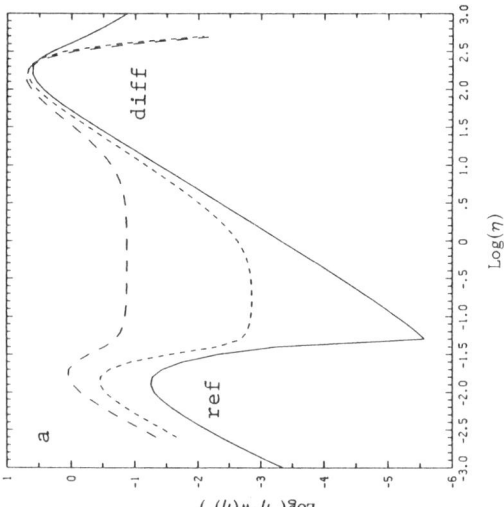

Figure 2. Taken from Romani, Narayan and Blandford (1986). η is the same as ql_F in Fig. 1 and $\tilde{w}(\eta)$ is the power spectrum of the two-dimensional spatial intensity fluctuations in a plane transverse to the line-of-sight at the observer. The refractive scale is given by $\log \eta_{\text{ref}} = -2.0$ and the diffractive scale by $\log \eta_{\text{diff}} = +2.0$. (a) The solid curve corresponds to a Type A phase spectrum with $\beta = 11/3$. The short- and long-dashed curves are for Type A1 spectra with the same β and with inner scales given by $q_{\max}l_F = 100\,\eta_{\text{ref}}$ and $10\,\eta_{\text{ref}}$ respectively. (b) The solid curve corresponds to a Type B spectrum with $\beta = 4.3$. The short- and long-dashed curves are for Type B1 spectra with the same β and with inner scales given by $q_{\max}l_F = 100\,\eta_{\text{ref}}$ and $10\,\eta_{\text{ref}}$ respectively.

spectrum with $\beta > 11/3$ will ease the problem with m_{ref}, though it may worsen the agreement in the $\Delta\nu_{\text{dc}}$ vs λ scaling.

Goodman and Narayan (1988, this volume, hereafter GN2) have considered the appearance of scatter-broadened images for various density fluctuation spectra. They find that Type A spectra should produce fairly smooth images, smoother than with the other spectra considered below. Recent VLBI observations of scatter-broadened radio sources appear to be compatible with this expectation (Wilkinson, Spencer and Nelson 1987, and several papers in this volume). The elliptical distortion of the image is expected to be small, ~ 0.1 (Romani, Narayan and Blandford 1986), which is again in agreement with the data. Observations of image motion as a function of time give another potential handle on the density fluctuation spectrum. For Type A spectra, the rms image wander, $\Delta\theta$, is given by

$$\Delta\theta \sim \theta_{\text{scatt}}(q_{\text{ref}}/q_{\text{diff}})^{(4-\beta)/2}. \qquad (7)$$

The work of Gwinn et al. (this volume) confirms that $\Delta\theta$ for Sgr B2 and several H_2O masers is consistent with $\beta = 11/3$ and inconsistent with $\beta \gtrsim 3.8$.

The dynamic scintillation spectrum of pulsars has proved to be another sensitive probe of the density fluctuation spectrum in the ISM. Patterns of drifting bands have been found to be common in the dynamic spectra of many pulsars. For Type A phase spectra, it is expected that the slope, $d\nu/dt$, of these bands would have a magnitude

$$\left|\frac{d\nu}{dt}\right| \sim \frac{\Delta\nu_{\text{dc}}}{t_{\text{scint}}}\left(\frac{q_{\text{ref}}}{q_{\text{diff}}}\right)^{(4-\beta)/2}. \qquad (8)$$

Smith and Wright (1985) measured drift slopes in several pulsars and found the data to be consistent with the slopes expected for $\beta = 11/3$. However, dynamic spectra frequently display periodic and/or criss-cross patterns and this, according to our present understanding, is inconsistent with a shallow Type A spectrum (Roberts and Ables 1982, Hewish, Wolszczan and Graham 1985). Some pulsars exhibit epochs of very sustained periodicity (e.g. Wolszczan and Cordes 1987), indicating that the pulsar image is temporarily dominated by a small number of bright spots (possibly just two spots). Such spotty images conflict with the smooth profile expected with a Type A spectrum.

To summarize, a Type A spectrum with a Kolmogorov index, $\beta = 11/3$, is consistent with a wide range of observations and must at this time be considered the front-runner. The strongest discrepancies are associated with (i) the enhanced level of refractive scintillation and (ii) the occasional occurrence of dramatic periodicities in pulsar dynamic spectra.

2.2 Type A1 Spectrum (Shallow Spectrum with Inner Scale)

The dashed lines in Fig. 2a show the intensity fluctuation spectra expected for a Type A1 phase spectrum with $\beta = 11/3$ and two different inner scales. Compared to a Type A spectrum, there are two differences:

(i) The refractive peak at q_{ref} is enhanced. It is this feature that led Rickett et al. (1984) originally to propose the Type A1 spectrum for the ISM. Coles et al. (1987) have pursued this aspect of a Type A1 spectrum in greater detail.

(ii) The valley between the diffractive and refractive peaks acquires extra power. Goodman et al. (1987) showed that this component in the intensity spectrum is due to caustics, which become particularly strong when the inner scale approaches the refractive scale.

Along with the enhanced refractive scintillation, the scatter-broadened image too becomes less smooth than with the Type A spectrum (GN2), thus providing a greater likelihood of explaining periodicities in dynamic spectra; however, as Goodman et al. (1987) argued, Type A1 spectra probably cannot explain some of the most dramatic instances of periodicities.

The strongest single difficulty with the Type A1 spectrum is that it predicts scalings in eq. (5) corresponding to $\beta = 4$. As mentioned earlier, the work of CWB rules out this scaling for many pulsars. This is not yet a fatal objection since the pulsars studied by CWB are not the ones with the strongest need for an A1 spectrum. More measurements of these scalings are needed.

2.3 Type B Spectrum (Steep Spectrum)

The solid line in Fig. 2b shows the intensity fluctuation spectrum for a Type B phase spectrum with $\beta = 4.3$. The most striking difference compared to the Type A case (Fig. 2a) is the strongly enhanced refractive peak. This aspect of Type B spectra was highlighted by Blandford and Narayan (1985) and GN and was the motivation behind the suggestion by these authors that the spectrum of density fluctuations in the ISM may be of Type B.

For a Type B spectrum the scalings in eq. (5) are replaced by (GN)

$$t_{\text{scint}} \propto \lambda^{-(\beta-2)/(6-\beta)}, \quad \Delta\nu_{\text{dc}} \propto \lambda^{-8/(6-\beta)}, \quad \theta_{\text{scatt}} \propto \lambda^{4/(6-\beta)}. \qquad (9)$$

GN showed that $\beta = 4.2$ in eq. (9) gives almost the same scalings as $\beta = 11/3$ in eq. (5), and so CWB's observations on the scaling of $\Delta\nu_{\text{dc}}$ are simultaneously consistent with both values of β. In addition, a Type B spectrum will frequently produce large image splittings, leading naturally to dramatic periodicities in dynamic spectra.

Despite these positive features, the Type B hypothesis is beginning to appear unlikely today. Probably, the strongest argument against it is that it predicts scatter-broadened images with substantial sub-structure (of a fractal nature, GN2). This translates to a strong power-law tail at long baselines in interferometric visibilities of scatter-broadened images. Wilkinson et al. (1987) looked for this in Cyg X-3 but found no supporting evidence. More evidence has been presented at this meeting to show that visibilities fall off exponentially.

Further, the work of GN2 suggests that scatter-broadened images should be split and periodicities should be seen in dynamic spectra very frequently (maybe half the time), but observations suggest that they occur only about 10% of the time. Thus, even while the Type B spectrum naturally explains periodic spectra by producing patchy images, it probably overdoes it significantly.

A Type B spectrum with large outer scale predicts much larger drift slopes in dynamic spectra than those observed. To bring the theoretical predictions into agreement with the observations of Smith and Wright (1985) it is necessary to postulate an outer scale $r_{ref} \lesssim 10^{14}$ cm. This may be in conflict with the detection of drifts in PSR 1937+214 (Cordes et al. 1986), which requires $r_{ref} > 10^{14}$ cm for this line-of-sight. Limits on r_{ref} can also be placed using image wander, which for a Type B spectrum is given by (compare with eq. 7)

$$\Delta\theta \sim \theta_{scatt}(q_{ref}/q_{min})^{(\beta-4)/2}. \tag{10}$$

2.4 Type B1 Spectrum (Steep Spectrum with Inner Scale)

The dashed lines in Fig. 2b show the intensity fluctuation spectra for Type B1 phase spectra. The diffractive and refractive peaks are similar to those with the Type B phase spectrum, but there is an additional component in the valley between the peaks due to caustics. Indeed, the Type B1 spectrum behaves almost identically to the Type B spectrum in all respects, except for the additional feature of displaying caustics. The close resemblance between the Type B and Type B1 spectrum was one of the arguments used by GN in support of steep spectra for the ISM. If (i) there are reasons to believe that the ISM cannot support density fluctuations on lengthscales as short as r_{diff}, and (ii) the scalings of t_{scint}, $\Delta\nu_{dc}$ and θ_{scatt} with λ differ from those predicted for $\beta = 4$ (as the CWB observations seem to indicate), then the B1 spectrum is the only consistent option within the family of power-law spectra. However, all the reasons for rejecting the Type B spectrum are equally (or even more) applicable to Type B1 spectra, and such spectra must now be considered less likely than spectra of Type A or Type A1.

2.5 Type A2, B2 Spectra (Spectra with Outer Scale)

These spectra have $r_{out} < r_{ref}$ and hence do not display refractive effects, i.e. in Fig. 2, their intensity spectra would have merely the diffractive peak at large wave-vectors. These phase spectra Types are therefore ruled out for those lines-of-sight that display strong refractive intensity variations or drifting bands and periodicities in dynamic spectra. However, there are highly-scattered lines-of-sight (e.g. Cygnus X-3, Sgr A) where the refractive timescales are so large (~ 100 y) that it is not yet possible to tell whether or not refractive effects are present. Such lines-of-sight might have Type A2 or B2 spectra, particularly if r_{out} is as small as $\sim 10^{14}$ cm. Note that the same physical spectrum could correspond to different classification Types at different scattering strengths (or λ); for instance, a medium may correspond to Type A at low levels of scattering, but could change to Type A1 or A2 (or even A12, which is like a single-scale spectrum) at higher levels of scattering, because r_{diff} becomes smaller with increasing scattering and r_{ref} becomes larger.

A Type A2 spectrum has diffractive scalings similar to those of the equivalent Type A spectrum (eq. 5), but the scalings of a Type B2 spectrum change from the corresponding Type B scalings of eq. (9) to those of a single-scale spectrum ($\beta = 4$ in eq. 5 or eq. 9).

2.6 Summary of Power-Law Spectra

Summarizing the previous five sub-sections, the leading contender is still the unconstrained Kolmogorov spectrum of Lee and Jokipii (1975), viz. Type A with $\beta = 11/3$. This spectrum has the best theoretical basis among all postulated spectra and has a substantial amount of supporting observational evidence. The spectrum fails in the following two counts:

(i) It predicts a refractive modulation index $m_{\text{ref}} \sim 0.1\text{--}0.2$, whereas several pulsars appear to vary at much higher levels, $m_{\text{ref}} \sim 0.5$. This is a serious discrepancy, but most of the observations are rather old and it is important to collect newer data with today's improved techniques. Also, the theoretical m_{ref} quoted above is for a thin screen, while an extended medium gives values of m_{ref} that are 2 to 3 times larger (Codona et al. 1986, Romani et al. 1986, Codona and Frehlich 1987). The extended medium is probably not a good model for lines-of-sight longer than a few kpc through the galactic plane (see sec. 3), but the pulsars with high values of m_{ref} are typically close to the sun and an extended medium may be appropriate for some of them.

(ii) The Type A Kolmogorov spectrum predicts smooth scatter-broadened images that apparently cannot produce the spectacular periodicities that are sometimes seen in pulsar dynamic spectra. This is again a serious discrepancy, but it should be remembered that the nature of scintillation in individual realizations of the scattering screen is not fully understood, and one cannot rule out the possibility that some level of periodicity may be produced occasionally even with a shallow spectrum. Also, an extended medium may cause important modifications that are not yet understood.

If one decides to explore power-law spectra other than the Type A Kolmogorov spectrum, there are three possibilities:

(i) Type A spectrum with $\beta > 11/3$: Such a spectrum will retain some advantages of the Kolmogorov spectrum and will alleviate the two failings discussed above; however, it will fit the CWB scaling data less well.

(ii) Type A1 spectrum with $\beta = 11/3$ and $r_{in} \gtrsim 10^{11}$ cm: This spectrum again is likely to reduce the above two discrepancies. The scalings of $\Delta\nu_{dc}$ with λ observed by CWB appear to rule out this spectrum but it would be preferable to carry out more observations before passing a final judgement.

(iii) Type B or B1 spectrum: These spectra will have no problems with the two failings of the Kolmogorov spectrum discussed above and may, in fact, overdo things in those respects. As discussed in section 2.3, the outer scale r_{out} needs to be $\lesssim 10^{14}$ cm. The strongest evidence against these spectra comes from VLBI observations which have revealed smooth scatter-broadened images, in strong contrast to the fractally-fragmented images predicted by GN2.

2.7 Other Spectra

One possibility to keep in mind is that the power spectrum of density fluctuations in the ISM need not be purely of power-law form. Very little work has been done on non-power-law spectra. Cordes and Wolszczan (1986) suggested that most of the observations, including the enhancement of refractive effects, could be explained by an extended Kolmogorov Type A spectrum with an additional "bump" of density fluctuation power at the refractive scale, $r_{ref} \sim 1$ AU.

The importance of the bump relative to the smaller scale fluctuations can be determined through Fig. 1a. If, despite the bump, the spectrum near q_{ref} lies below the dashed line, the scattering is still dominated by power near q_{diff} and the bump may be treated as a perturbation. This is probably what Cordes and Wolszczan (1986) had in mind since it retains all of the characteristics of the Kolmogorov Type A spectrum but provides enhanced refractive effects. However, if the bump causes the spectrum near q_{ref} to lie above the dashed line, then the small scales become irrelevant. The spectrum would then become essentially of Type B1 and will display violent caustics, spotty images, etc. As mentioned earlier, this is ruled out as a model for the bulk of the ISM. However, Cordes and Wolszczan (1986) and Wolszczan and Cordes (1987) do report occasional episodes when the image of a pulsar consists of two sub-images separated by at least twice their individual sizes, which implies a spectrum that is close to Type B1 during these epochs.

An interesting possibility is that the bump in the spectrum represents a sporadic component of density fluctuations that occurs randomly only in some lines-of-sight. When present, the bump would cause large intensity fluctuations, wide image splittings and striking periodicities in dynamic spectra, but the effects would be absent most of the time. Support for this possibility may be found in the extreme scattering events discovered by Fiedler et al. (1987) in some extragalactic sources, which are apparently caused by lumps of size

~1 AU in the ISM. An interpretation of these events in terms of caustics has been given by Romani, Blandford and Cordes (1987).

3. DISTRIBUTION OF SCATTERING IN THE ISM

The normalization of the density fluctuation spectrum is described by the parameter, C_N^2, in eq. (1). In presenting observational results, it is usual to assume a power-law index $\beta = 11/3$, and to express C_N^2 in the units $m^{-20/3}$; we employ this convention in the numerical values quoted below. Note that for a general power-law with arbitrary β, C_N^2 takes on units of $m^{-(3+\beta)}$, which makes it difficult to compare values of C_N^2 corresponding to different β. Since all models of the density spectrum are agreed that there are fluctuations on scales ~1 AU, it may be best to express C_N^2 in the mixed units $m^{-6}\,AU^{(3-\beta)}$, which will facilitate comparison of different spectra. Another point to note is that estimates of C_N^2 in the ISM are usually obtained under the assumption of a uniform homogeneous medium, extending either all the way to the source in the case of galactic sources, or extending to the edge of the Galaxy (defined to be at some standard radius) for extragalactic sources. Since the medium is actually quite clumpy, individual regions of the ISM must have levels of C_N^2 much in excess of the estimated mean.

The most comprehensive analysis of the distribution of density fluctuations in the ISM is due to CWB. Based on measurements of scintillation parameters of a large number of pulsars, these authors concluded that the scattering in the ISM arises from two distinct components:

(i) There is a strongly-scattering component in the galactic plane, concentrated in the inner Galaxy, with scale height ≤100 pc, and mean $\log C_N^2 \sim 0$. This component is clumpy, with a mean spacing ~1 kpc between clumps.

(ii) In addition, there is a moderately-scattering nearly-uniform component with $\log C_N^2 \sim -3.5$ and scale height $\geq$ 500 pc.

For lines-of-sight longer than ~ 1 kpc in the plane of the Galaxy, the scattering by the clumpy component tends to dominate over that due to the extended component and in such cases the thin screen theory of scintillation is probably a good approximation. Lines-of-sight to nearby sources and sources at high latitudes probably sample only the extended medium and in these cases the full formalism of extended scattering may be necessary.

There is no unambiguous identification yet as to which phase of the ISM produces the clumpy component of scattering. There is evidence that this scattering may be due to HII regions. The Vela pulsar, for instance, has an extraordinarily high $\log C_N^2 \sim 0$, and most of its scattering appears to be caused by the Gum Nebula, identified as an HII region (e.g. Reynolds 1976). Also, the inner regions of the Galaxy are known to be the sites of giant HII regions, particularly in the 5 kpc ring. The scale height of these HII regions and the covering factor and electron densities of their low-density envelopes are consistent with the extraordinarily high level of $\log C_N^2$ inferred for this region of the Galaxy (see Anantharamaiah and Narayan, this volume).

There is other evidence to suggest that non-thermal phases of the ISM too may contribute to the clumpy scattering. Shapirovskaya (1978) pointed out that spurs, loops and ridges in the radio continuum map of the Galaxy are sites where compact background radio sources exhibit the strongest low-frequency variability. Rickett et al. (1984) have convincingly identified this variability with refractive scintillation. The enhanced scintillation of these sources probably implies enhanced scattering along their lines-of-sight. Heeschen et al. (1987) confirmed Shapirovskaya's result for short timescale flicker at higher radio frequencies. Also, Hjellming and Narayan (1986) have made a strong case for enhanced scattering in the extragalactic source, 1741-038, and identified the source of scattering to be the North Polar Spur, an old supernova remnant.

Meanwhile, the extreme scattering events discovered by Fiedler et al. (1987) have introduced a new component into the puzzle. Blobs of electron density with sizes ~ 1 AU and projected surface densities $\sim 10^{17}\,\text{cm}^{-2}$ are apparently abundant in the ISM, but their origin, location and lifetimes are yet unknown. As mentioned in section 2.7, these blobs may play an important role in producing some of the features of scintillation, like periodic dynamic spectra, that are currently unexplained.

The uniform moderately-scattering component (ii) of CWB is probably located in the general ionized ISM, which has a mean electron density of ~ 0.025 cm^{-3} (e.g., Lyne, Manchester and Taylor 1985) and a scale height ~ 1 kpc (e.g., Vivekanand and Narayan 1982).

Acknowledgement: I gratefully thank R. D. Blandford, J. M. Cordes, J. J. Goodman, J. R. Jokipii and R. W. Romani for reading the manuscript, suggesting several improvements, and generally ensuring that this review is fair and objective. This work was supported in part by the National Science Foundation under grant AST-8611121.

REFERENCES

Armstrong, J. W., Cordes, J. M., and Rickett, B. J. (1981). *Nature*, **291**, 561.
Backer, D.C. (1974). *Astrophys. J.*, **190**, 667.
Blandford, R., and Narayan, R. (1985). *Mon. Not. R. Astron. Soc.*, **213**, 591.
Blandford, R., Narayan, R., and Romani, R. W. (1986). *Astrophys. J. Lett.*, **301**, L53.
Cesarsky, C. J. (1980). *Ann. Rev. Astron. Astrophys.*, **18**, 289.
Codona, J. L., Creamer, D. B., Flatte, S. M., Frehlich, R. G., and Henyey, F. S. (1986). *Radio Sci.*, **21**, 805.
Codona, J. L., and Frehlich, R. G. (1987). *Radio Sci.*, **22**, 469, 481

Coles, W. A., Frehlich, R. G., Rickett, B. J., and Codona, J. L. (1987). *Astrophys. J.*, **315**, 666.
Cordes, J. M., Pidwerbetsky, A., and Lovelace, R. V. E. (1986). *Astrophys. J.*, **310**, 737.
Cordes, J. M., Weisberg, J., and Boriakoff, V. (1985). *Astrophys. J.*, **288**, 221 (CWB).
Cordes, J. M., and Wolszczan, A. (1986). *Astrophys. J. Lett.*, **307**, L27.
Fiedler, R. L., Dennison, B., Johnston, K. J., and Hewish, A. (1987). *Nature*, **326**, 675.
Fyfe, D., Joyce, G., and Montgomery, D. (1977). *J. Plasma Phys.*, **17**, 317.
Gochelashvily, K. S., and Shishov, V. I. (1975). *Opt. Quant. Electron.*, **7**, 524.
Goodman, J., and Narayan, R. (1985). *Mon. Not. R. Astron. Soc.*, **214**, 519 (GN).
Goodman, J., and Narayan, R. (1988). This volume (GN2).
Goodman, J., Romani, R. W., Blandford, R. D., and Narayan, R. (1987). *Mon. Not. R. Astron. Soc.*, **229**, 73.
Heeschen, D. S,., Krichbaum, Th., Schalinsky, C. J., and Witzel, A. (1987). *Astron. J.*, **94**, 1493.
Hewish, A., Wolszczan, A., and Graham, D. A. (1985). *Mon. Not. R. Astron. Soc.*, **213**, 167.
Higdon, J. C. (1984). *Astrophys. J.*, **285**, 109.
Hjellming, R. M., and Narayan, R. (1986). *Astrophys. J.*, **310**, 768.
Lee, L. C., and Jokipii, J. R. (1975). *Astrophys. J.*, **201**, 532.
Lee, L. C., and Jokipii, J. R. (1976). *Astrophys. J.*, **206**, 735.
Lo, K. Y., Backer, D. C., Ekers, R. D., Kellermann, K. I., Reid, M., and Moran, J. M. (1985). *Nature*, **315**, 124.
Lovelace, R. V. E. (1970). Ph.D. thesis, Cornell University.
Lyne, A. G., Manchester, R. N., and Taylor, J. H. (1985). *Mon. Not. R. Astron. Soc.*, **213**, 613.
Narayan, R., and Goodman, J. (1988). *Mon. Not. R. Astron. Soc.*, submitted.
Narayan, R., and Hubbard, W. B. (1988). *Astrophys. J.*, **325**, 503.
Reynolds, R. J. (1976). *Astrophys. J.*, **203**, 151.
Rickett, B. J. (1977). *Ann. Rev. Astron. Astrophys.*, **15**, 479
Rickett, B. J. (1986). *Astrophys. J.*, **307**, J564.
Rickett, B. J., Coles, W. A., and Bourgois, G. (1984). *Astron. Astrophys.*, **134**, 390.
Roberts, J. A., and Ables, J. G. (1982). *Mon. Not. R. Astron. Soc.*, **201**, 1119.
Romani, R. W., Blandford, R. D., and Cordes, J. M. (1987). *Nature*, **328**, 324.
Romani, R. W., Narayan, R., and Blandford, R. D. (1986). *Mon. Not. R. Astron. Soc.*, **220**, 19.
Scheuer, P. A. G. (1968). *Nature*, **218**, 920.
Shapirovsky, N. Ya. (1978). *Sov. Astron.*, **22**, 544.
Smith, F. G., and Wright, N. C. (1985). *Mon. Not. R. Astron. Soc.*, **214**, 97.
Vivekanand, M. and Narayan, R. (1982). *J. Astrophys. Astron.*, **3**, 399.
Wilkinson, P. N., Spencer, R. E., and Nelson, R. F. (1987). IAU Colloquium 129: The Impact of VLBI on Astrophysics and Geophysics.
Wolszczan, A., and Cordes, J. M. (1987). *Astrophys. J. Lett.*, **320**, L35.

OBSERVATIONS OF DIFFRACTIVE INTERSTELLAR SCINTILLATION PHENOMENA

Steven R. Spangler
University of Iowa, Iowa City, Iowa, 52242

ABSTRACT

Recent observations of radio source angular broadening and pulsar dynamic spectra are reviewed, with the goal of determining the spectrum of the interstellar density irregularities and the sources which generate the turbulence. Measurements of angular broadening and certain aspects of pulsar dynamic spectra favor "quasi-Kolmogoroff" density spectra, which are power law with an index less than four. However, pulsar dynamic spectra frequently display phenomena such as drifting bands or quasi-periodic structures which seem to require spectra with more power at long wavenumbers than can be supplied by quasi-Kolmogoroff spectra. The resolution of this seemingly contradictory situation is not obvious at the present time. Observations of angular broadening of extragalactic sources and spectral corrugation of pulsars show that the intensity of turbulence in the interstellar medium must be highly spatially inhomogeneous. The paper concludes with suggestions for future observations.

INTRODUCTION AND DEFINITION

This paper will summarize observations of diffractive interstellar scintillations, and discuss their implication for the nature of plasma turbulence in the interstellar medium. At the outset a point of nomenclature is in order, regarding the definition of diffractive scintillation phenomena. It might seem artificial to distinguish between refractive and diffractive phenomena, when the properties of different types of measurements, such as the angular broadening of a point source viewed through the medium, vary continuously as the form of the density spectrum is changed.

I will introduce two definitions, an empirical one and a physical one. According to the empirical definition, we may define diffractive scintillations as the phenomenon of *angular broadening* of any type of radio source, and *spectral corrugation* of radiation from pulsars. The term spectral corrugation will be used to describe the modulation of radiation in frequency and space, as revealed in a dynamic spectrum.

Beyond this minimalist empirical definition is a physical one, which motivated the distinction between refraction and diffraction in the first place. This definition is illustrated in Figure 1. Diffractive phenomena are dominated or exclusively produced by the short-scale phase fluctuations, ϕ_d, which in turn are engendered by the small scale density fluctuations in the medium. Refractive effects are produced by ϕ_r.

As we will see later in this paper, the validity of the physical definition actually depends on the form of the density power spectrum.

PLAN OF THE PAPER

The goal of the research discussed at this meeting is to understand the nature and generating mechanisms of the interstellar turbulence. The contribution of observational radio astronomy is to determine the spatial spectrum of the density irregularities, and

Scintillation Theory

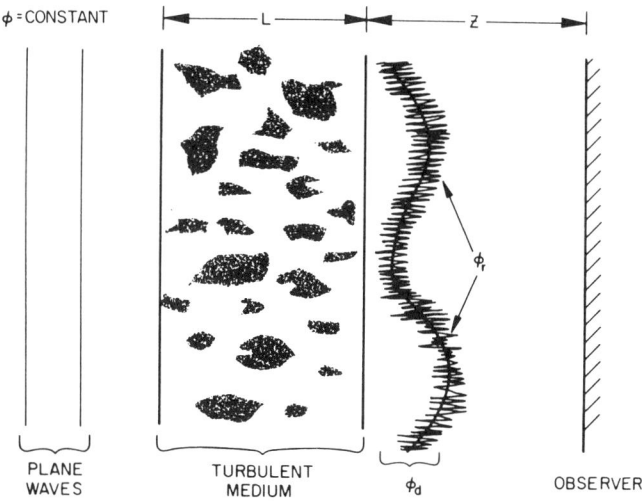

Figure 1: Wave propagation through a thick screen possessing random irregularities. Phase fluctuations are of two kinds: ϕ_d occurring on short spatial scales, and ϕ_r occurring on large spatial scales. Diffractive phenomena are produced by ϕ_d.

the galactic distribution of the turbulence. Identification of intense turbulence with specific types of astronomical objects would be particularly helpful in the second regard.

Previous observational projects have shown that the spectrum is roughly of the form

$$P_{\delta n}(\vec{q}) = C_N^2 q^{-\alpha} \qquad (1)$$

with $\alpha \sim 4$. By determination of the spectrum, we now mean specifying α in the interval 3.5 to 4.5, or elucidating equivalent modifications to (1).

There are two reasons why such an enterprise is not simply latter day scholasticism.

- Determination of the spectrum provides a clue to the mechanisms generating the turbulence. The precise value of the slope is perhaps not so crucial as the resultant implications about the inner and outer scales of the turbulence.

- The value of the index α within the aforementioned range determines whether interstellar scintillation is responsible for a set of other phenomena, most notably the low frequency variability of extragalactic radio sources.

A few remarks are appropriate concerning the physical interpretation of the spectrum (1). It is generally felt that if α is close to 3.67, the "Kolmogoroff" value, then the spectrum is probably power law over a wide wavenumber range, since physical arguments for such a spectrum have been advanced[1]. Similar arguments have not been made for spectra with α substantially in excess of 3.7. If observations infer that α is much larger than 3.7, it is possible that we are using a large α to parameterize a more complicated density spectrum. For example, we might have a Kolmogoroff spectrum with additional, possibly nonrandom structures such as shock waves at low wavenumbers. Other possible modifications have been described by Coles and colleagues[2]. For the remainder of this talk, we will refer to spectra with $\alpha < 4$ as "quasi-Kolmogoroff", while those with α inferred to be in excess of 4 will be termed simply "steep spectra".

ANGULAR BROADENING

Angular broadening is the scattering or blurring of an image viewed through a turbulent medium. It is represented mathematically by the two point correlation function[3],

$$< E(\vec{x_1})E(\vec{x_2}) >= V(\vec{x_1} - \vec{x_2})\exp(-\frac{1}{2}D_\phi(r)) \qquad (2)$$

where

$$D_\phi(r) = 8\pi^2 r_e^2 \lambda^2 \int_0^L dz \int_0^\infty dq q(1 - J_0(qr))P_{\delta n}(q,z) \qquad (3)$$

where $r \equiv |\vec{x_1} - \vec{x_2}|$ is the distance between antennae, and isotropic turbulence is assumed. The thickness of the scattering medium is given by L, and $V(\vec{x_1} - \vec{x_2})$ is the intrinsic correlation function of the source of radiation, which is the interferometric visibility.

Equations (2) and (3) may be used to demonstrate the ambiguity regarding the size of irregularities responsible for diffractive phenomena. Let us imagine that a VLBI interferometer with a baseline $\vec{r} = \vec{x_1} - \vec{x_2}$ measures a visibility of V_0 on some source. For homogeneous turbulence equations (2) and (3) give us

$$8\pi^2 r_e^2 \lambda^2 L \int_0^\infty dq q(1 - J_0(qr))P_{\delta n}(q) = -2\ln(\frac{V_0}{V(\vec{r})}) \qquad (4)$$

Let us now substitute equation (1) into (4), and integrate between the limits q_0 and q_1, which are the inner and outer scales of the turbulence. We also assume, for purposes for illustration, that $qr \ll 1$, which means that the irregularities are much larger than the interferometer baseline. Equation (4) then becomes

$$\frac{2\pi^2 r_e^2 \lambda^2 L C_N^2 r^2}{(4-\alpha)}(q_1^{4-\alpha} - q_0^{4-\alpha}) = -2\ln(\frac{V_0}{V(\vec{r})}) \qquad (5)$$

Equation (5) demonstrates the significance of the spectral index α, which determines whether the smallest irregularities (q_1^{-1}) or the largest (q_0^{-1}) determine our interferometric measurement. If $\alpha < 4$, the effects of density irregularities as small as the separation of the antennae, or the inner scale if it is larger, are measured. If $\alpha > 4$, the measured electric field correlation length is determined by large scale irregularities, and is in no way a measure of the correlation length of the density fluctuations in the medium, a point emphasized by Uscinski[4] some time ago.

Equations (4) and (5) illustrate some additional important points regarding the interpretation of angular broadening measurements.

- If the spectrum of interstellar density fluctuations is a power law with $\alpha > 4$, the observed visibility (corrected for intrinsic structure, if any) will be a Gaussian function of the baseline length, provided that the conditions for measurement of the ensemble average are met.

Scintillation Theory

- If the spectral index $\alpha < 4$, but the inner scale is much greater than the baseline length, then the observed visibility will be of the same form as that described immediately above, but the coefficient of the quadratic term in the argument will have a different significance.

- If the spectrum is quasi-Kolmogoroff, *and* the inner scale is smaller than the baseline length, then the logarithm of the visibility will be proportional to the baseline length to the power $\alpha - 2$.

These considerations indicate that angular broadening measurements can be extremely informative about the turbulence in the interstellar medium.

It may be argued that angular broadening is the optimum technique for studying interstellar turbulence. First of all, the observable, the two point electric field correlation function, is precisely what is measured by a radio interferometer. The substantial expansion and improvement of radio interferometers during the last decade has given us powerful tools for this application. Second, angular broadening affects all sources viewed through the medium. Angular broadening measurements have been made on extragalactic radio sources, sources of molecular maser radiation, and pulsars. Other diffractive scintillation phenomena are observable only for pulsars because of scintillation quenching by finite angular size. Finally, the theory of the two point correlation function is the best understood. Equation (2) is correct for extended or screen media, and for any irregularity spectrum. By way of contrast, the theory for spectral corrugation (specifically the *intensity* two point correlation function at different frequencies as well as different spatial locations) to be discussed later admits analytic solutions only under restrictive approximations, such as a thin screen, or a Gaussian irregularity spectrum.

An important caveat to be noted in the use of equations (2) and (3) is they refer to an ensemble average measurement. In contrast, the observations to which they are compared represent a temporal average, as well as other integrations resulting from the measurement process, such as the use of a finite bandwidth. At the present time we do not in general know the conditions necessary for the measured visibility to be identical to the ensemble average. If the irregularity spectrum is quasi-Kolmogoroff, then it appears that typical VLBI experiments will yield the ensemble average visibility[5]. However, for steep density spectra this may not be the case, and the measured visibility might depart substantially from that predicted by equations (2) and (3). This matter is discussed further in the paper by Dr. Goodman.

Of the types of radio sources which can be used in angular broadening measurements, pulsars are preferable. The reason for this is simply that they are so compact that $V(\vec{x_1} - \vec{x_2}) = 1$ for all achievable baselines and frequencies. Source structure does not contaminate the measurements, and properties of the interstellar turbulence can be inferred with greater confidence. Nonetheless, these objects have disadvantages as well. Pulsars lie at poorly known distances in the interstellar medium. Aside from the obvious geometric problem of not knowing which regions of galactic ionized hydrogen are interposed on the line of sight, differing distances between the pulsar and a dominant scattering region can drastically affect the strength and properties of the angular broadening.

Of more importance, radio interferometer observations of pulsars are challenging for technical reasons, and the intrinsic variability of pulsars can complicate the analysis of interferometer data; standard techniques of aperture synthesis assume a constant source of radiation. For these reasons, the limited measurements of pulsar angular broadening to date have not provided the most significant results emergent from this technique. This situation is in the process of changing, and the paper by Dr. Gwinn reports on measurements of pulsar scattering.

Existing radio interferometers were largely designed to study the structure of extragalactic radio sources; these objects would therefore seem to constitute poor sources of radiation for angular broadening experiments. However, it can be argued that research in extragalactic astronomy has provided considerable knowledge about these objects, hopefully sufficient to estimate $V(\vec{x}_1 - \vec{x}_2)$ for some objects, and assess the effect of intrinsic structure on the observed visibilities. If this can be done, the advantages of extragalactic sources are considerable. The observations are straightforward, as is the reduction of the data. The broadband nature of the objects means measurements can be made at a number of frequencies. The large number of extragalactic radio sources allows many lines of sight through the galaxy to be probed. If the Mark III recording system is used, the density of observable sources is so high that the vicinity of virtually any astronomical object of interest may be investigated. Finally, we know that for all extragalactic sources the line of sight passes through the entire interstellar medium, which simplifies the interpretation of the measurements.

A cynic could contend that sources of molecular maser radiation conjoin the undesirable features of pulsars and extragalactic radio sources, and append an additional one, monochromaticity. Nonetheless, their low galactic latitude means that they often probe regions of prominent scattering, and observations of maser angular broadening have contributed substantially to the field.

We will now describe three quite recent observational results which provide a measurement of the spectrum of interstellar density turbulence. The contribution of angular broadening measurements to our knowledge of the galactic distribution of turbulence will be deferred.

Spangler and Cordes[5] have carried out VLBI observations of the source 2013+370 using the U.S. network. Previous multifrequency observations demonstrated that its structure at frequencies below 5 GHz is dominated by interstellar scattering. Figure 2 shows a contour plot representation of the source brightness distribution. A least squares fit to the visibility yields a value of $\alpha = 3.79 \pm 0.05$. A similar type of observation was reported by Wilkinson and colleagues[6], who made observations of Cygnus X-3 at 408 MHz with the Merlin synthesis telescope. They chose to analyse times when the source was in a flaring state, and therefore presumably most compact, so that intrinsic structure would make a negligible contribution to the observed visibility. Wilkinson *et al* report a best fit value of $\alpha = 3.85 \pm 0.05$. The two angular broadening measurements are therefore in excellent agreement, even though they are obtained with different radio sources, interferometer arrays, and frequencies of observation.

An additional recent project in the angular broadening *gens* is that of Gwinn and colleagues[7]. Observations in three epochs were made of H_2O (22 GHz) masers in Sgr B2(N). The measured size distribution for the maser spots was truncated at about 300 microarcseconds, which Gwinn *et al* identify as the diffractive scattering size. Large scale gradients in the wave phase, indicated by ϕ_r in Figure 1, cause position wander on the sky: a manifestly refractive effect. As will be discussed below, the ratio of the

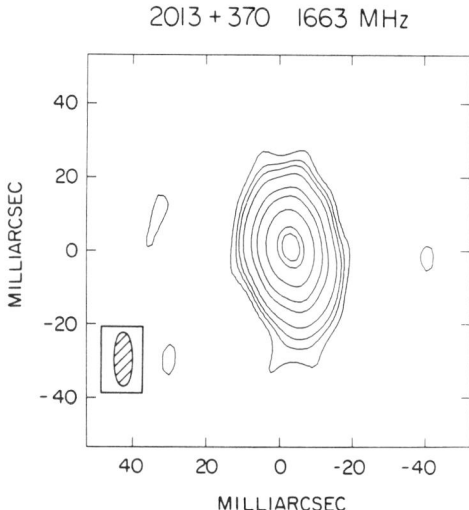

Figure 2: Scattering image of the radio source 2013+370 at 1663 MHz (Spangler and Cordes (1988)). A least squares fit to the visibility function yields a value for the electron density spectral index, $\alpha = 3.79 \pm 0.05$. Contours are at 2,4,6,10,20,30,50,80, and 90 percent of the peak intensity.

refractive angular offset, θ_r to the diffractive scattering size, θ_d is a strong function of the irregularity spectral index α. For steep spectra, θ_r is comparable to or larger than θ_d. Thus if we were observing the Sgr B2 masers through a turbulent medium with a steep irregularity spectrum, we would expect random peregrinations of the maser features, with the refractive excursions being comparable to or larger than the diffractive scattering size.

Gwinn et al place a limit of approximately 20 microarcseconds on the magnitude of such refractive wander. This limit is obviously much smaller than the diffractive scattering size, and so the observations favor a quasi-Kolmogoroff spectrum. An analysis indicates that the observations are inconsistent with spectral indices much greater than about 3.7.

In conclusion, three independent angular broadening measurements are in consensus that the mean spectrum of irregularities is quasi-Kolmogoroff, and not steep enough to produce pronounced refractive effects.

PULSAR SPECTRAL MEASUREMENTS

In this section we consider phenomena observed in the dynamic spectra of pulsars. To provide a focus for discussion, Figure 3 shows an example of such a spectrum. The theoretical quantity conjugate to such data is the intensity two point correlation function at different frequencies and spatial locations. In the limit of strong scattering, appropriate to interstellar scattering at the frequencies we are considering, this correlation function reduces to the square of the *electric field* two point correlation function

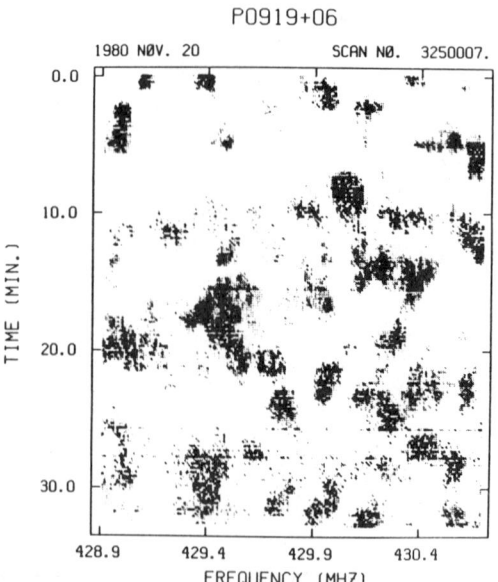

Figure 3: Example of an observed pulsar dynamic spectrum. Correlation functions of data such as this, with lags in time and frequency, have formed the basis for discussions of the interstellar density spectrum. The observation is by J.M. Cordes, J.M. Weisberg, and V. Boriakoff.

at different frequencies[8]. The observable quantity to which this theoretical correlation is referred is the correlation function of the pulsar spectrum, with the lag being in frequency, time of observation, or both.

It would be gratifying if the analysis of pulsar dynamic spectra confirmed the conclusions of the previous section, i.e. that the density spectrum in the interstellar medium is quasi-Kolmogoroff. Indeed, such evidence is not absent.

If one considers a horizontal slice in Figure 3, a characteristic frequency would emerge over which the intensity is correlated. This is referred to as the decorrelation bandwidth. The decorrelation bandwidth depends on frequency in a way which is determined by the irregularity spectrum. Specifically,

$$B \propto \nu^{\frac{2\alpha}{\alpha-2}} \qquad (6)$$

where B is the decorrelation bandwidth and ν is the frequency of observation[3]. Equation (6) is valid for α in the range 2 to 4. Measurement of the frequency scaling of the decorrelation bandwidth therefore provides a measure of α. Cordes, Weisberg, and Boriakoff[3] utilized their own measurements as well as those in the literature to determine the frequency scaling of the decorrelation bandwidth for five pulsars. The mean inferred value of α was 3.63 ± 0.2, supporting a quasi-Kolmogoroff density spectrum.

Further evidence for a quasi-Kolmogoroff spectrum was provided by analysis of the shape of the frequency correlation function. Consider the autocorrelation function (with lag in frequency and at fixed time) of the spectrum shown in Figure 3. The resultant

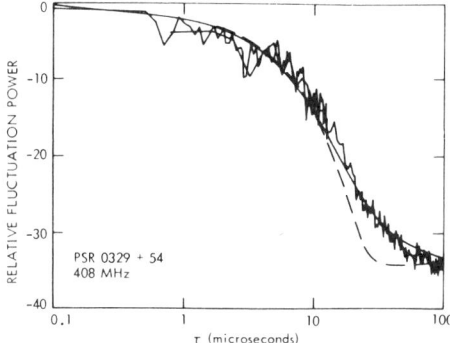

Figure 4: Observations of the "fluctuation power spectrum" for the pulsar 0329+54 from Armstrong and Rickett[8]. The function is the Fourier transform of the frequency autocorrelation function. Overplotted on the data are curves representing a Gaussian spectrum (dashed line) and a quasi-Kolmogoroff one with $\alpha = 3.9$ (solid line).

shape contains information on the spectrum of irregularities, in the same way that the visibility of a scattered source is determined by the irregularity spectrum (equations (2) and (3)).

Such an analysis has been carried out by Armstrong and Rickett[8] and Wolszczan[9], with consistent results. An illustration of the method and results is shown in Figure 4, which displays the data of Armstrong and Rickett. Plotted is a quantity termed $\tilde{\gamma}_4(\tau)$ by those authors, which is the temporal Fourier transform of the frequency autocorrelation function. Overplotted on the data are the expected relationships for a Gaussian density spectrum, equivalent to that produced by spectra with $\alpha > 4$, and a quasi-Kolmogoroff spectrum with $\alpha = 3.9$. The quasi-Kolmogoroff spectrum is clearly favored by the data. Similar results were obtained[8,9] for all pulsars possessing sufficient signal-to-noise ratio to discriminate between the two spectra; quasi-Kolmogoroff spectra with α between 3.8 and 3.9 were favored. It should be noted that the aforementioned admonitions concerning measured quantities and ensemble averages apply here as well.

A final argument in favor of quasi-Kolmogoroff density spectra has been raised by Cordes, Pidwerbetsky, and Lovelace[10], and though perhaps difficult to evaluate quantitatively, is intriguing to contemplate. As discussed above, the small scales are dominant for quasi-Kolmogoroff turbulence, whereas for steeper spectra large scales determine the observations. The net effect is that the time over which one must form the observation, in order to obtain the ensemble average, is much longer for steep spectra than quasi-Kolmogoroff spectra. Given the practical limitations on radioastronomical measurements, an observational average is unlikely to correspond to the theoretical ensemble average if the true turbulence spectrum is steep. The fact that the observed correlation functions even remotely resemble the theoretical ones strongly suggests that the spectra are quasi-Kolmogoroff.

The results from pulsar spectra discussed in the previous subsection are in complete accord with those inferred from angular broadening measurements. However, to terminate the discussion here would leave an entirely defective impression, namely that spectral corrugation observations are supportive of quasi-Kolmogoroff spectra, *in toto*.

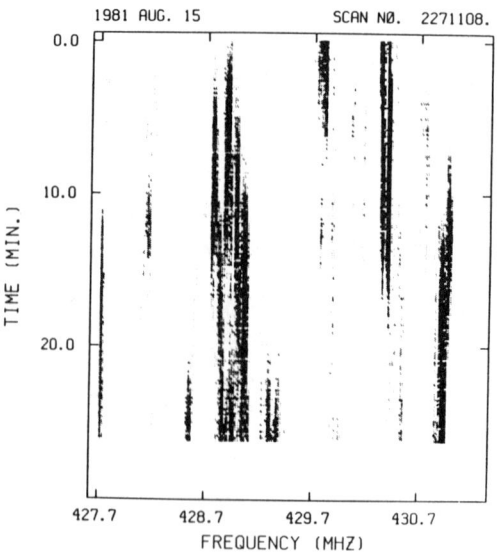

Figure 5: Dynamic spectrum of the pulsar 2016+28, showing drifting band structure.

In fact, perhaps the most compelling evidence for prominent *refractive* scintillation phenomena, caused by steep density spectra, are to be found in pulsar dynamic spectra.

The dynamic spectrum shown in Figure 3 illustrates a totally random process, with frequency and time being independent variables. Spectra of this sort were analysed by Armstrong and Rickett, with the conclusions stated in the previous subsection. From the beginning of the systematic study of pulsar scintillations however, it was realized that dynamic spectra are often much more organized. Frequently drift bands are observed in pulsar spectra. An example is shown in Figure 5. A more extreme example of organization in dynamic spectra is shown in Figure 6. Here the bands are arranged in a quasi-periodic structure. Viewing these dynamic spectra, one has the impression that a deterministic process is modulating or even supplanting the one responsible for the spectrum shown in Figure 3.

In 1980, Hewish[11], elaborating earlier suggestions of Shishov[12], discussed what is now widely considered the correct interpretation of the phenomena shown in Figures 5 and 6. The physical basis of the mechanism may be understood by reference to Figure 1. Since the vector wavenumber of a wave is the gradient of the phase, the large scale refractive phase changes ϕ_r cause shifts in the position of a source viewed through the medium. This is the same effect discussed above with respect to the observations of Gwinn *et al.* Hewish pointed out that since the interstellar medium is dispersive, the random diffraction pattern at the observer's screen will be shifted as the observing frequency is changed. In the case of net motion between the pulsar and the screen, caused in the interstellar medium by the space velocity of the pulsar, maxima and minima at different frequencies will move past the observer at different times, and bands such as those shown in Figure 5 will be observed.

The quasi-periodic structures shown in Figure 6 are interpreted as a more extreme case of the same process. Here the refractive tilting of the wave front causes two or more

Scintillation Theory 41

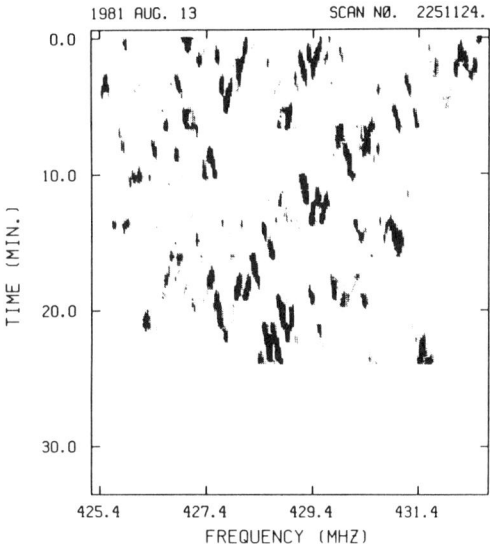

Figure 6: Dynamic spectrum of the pulsar 1919+21, showing quasi-periodic bands.

diffractive "beams" to constructively and destructively interfere at the observer's screen. This realization has led to the employment of this phenomenon as an interferometer of acute resolving power[13].

These features then are produced by large scale density fluctuations in the interstellar medium. In the mathematical model of scattering by a screen, the relevant scales are those comparable to or larger than the "multipath scale" $l_{mp} \equiv \theta_d z$, where z is the distance from the observer to the screen. For the case of quasi-Kolmogoroff spectra, this size is considerably larger than that of the irregularities responsible for the scattering disk.

A number of studies, of varying degrees of exactitude[10,11,14,15,16], have attempted to deduce the consequences of these features for the spectrum of density irregularities. There seems universal consensus that the common observation of quasi-periodic structures can result only from steep density spectra.

There appears to be less agreement on the implications of the drifting bands. Smith and Wright[17] measured the rate of drift of these features, $\frac{d\nu}{dt}$. For screen models of the turbulence, the frequency drift is given by[11,17]

$$\frac{d\nu}{dt} = \frac{\nu V \sec\phi}{\theta_r z} \qquad (7)$$

where V is the relative speed of the source and screen, ϕ is the angle between the net velocity of the source and screen and the gradient of the refractive wedge, and z is the pathlength through the medium. In the analysis of Smith and Wright, V was taken to be a "scintillation pattern speed", determined by the characteristics of the scintillations.

Measurement of the decorrelation bandwidth provided an estimate of the diffractive scattering angle θ_d, so that the ratio $\frac{\theta_r}{\theta_d}$ was estimated. Smith and Wright[17] concluded that the observed value, $\simeq 0.25$, was consistent with a quasi-Kolmogoroff spectrum.

However, while the frequency drift rates seem to be compatible with quasi Kolmogoroff irregularities, the shapes of the bands do not, as noted by Romani, Narayan, and Blandford[16]. The same theory[10] that yields equation (7) also predicts that the width to length ratio of the bands should be $\sim \frac{4\theta_r}{\theta_d}$. For a quasi-Kolmogoroff spectrum, we would then expect the bands to be only slightly elongated circles, in striking contrast to the observed phenomena, such as Figure 5. Further evidence that something is amiss with the drift rate data may be inferred from the data of Cordes and Wolszczan[13]. These authors computed the refractive angle θ_r in a method equivalent to that of Smith and Wright[17], and found $\theta_r < \theta_d$ for the pulsar 0919+06, even when the pulsar was displaying quasiperiodic features. As mentioned above, quasiperiodic features are a patently refractive phenomenon, so the estimation of the refractive angle from frequency drifts seems suspect. We may plausibly conclude that the drifting bands, as well as the quasi-periodic features, require the agency of a steep density spectrum in the interstellar medium.

In addition to these positive arguments for steep spectra, some authors have raised objections to the evidence in favor of quasi-Kolmogoroff spectra given in the previous subsection. Goodman and Narayan[18] and Cordes, Pidwerbetsky, and Lovelace[10] have pointed out that the frequency scaling of the decorrelation bandwidth is not an unambiguous measure of α. From Figure 16 of Cordes, Pidwerbetsky, and Lovelace[10], or Figure 4 of Goodman and Narayan[18], we see that the spectral index of the decorrelation bandwidth is a double valued function of α. A value of $\alpha = 4.15$ is as acceptable as the quoted value of 3.63 in interpreting the decorrelation bandwidth data of Cordes, Weisberg, and Boriakoff[3]. Furthermore, Hewish, Wolszczan, and Graham[15] have speculated as to whether the correlation functions of the sort shown in Figure 4 truly satisfy the conditions for measurement of an ensemble average, and therefore whether these measurements can be used to exclude steep density spectra.

In concluding this subsection, it is necessary to emphasize that the observed spectra are highly variable. The spectral character observed for a given pulsar can vary from one observing session to another. This is amply illustrated in Figure 1 of Cordes and Wolszczan[13], which shows dynamic spectra, at the same observing frequency, of the pulsar 0919+06 during three observing sessions. One sees a transition from a completely random spectrum to quasi-periodic features and then to drifting bands. This feature of the observed behavior greatly complicates attempts to determine the type of density spectrum affecting the pulsar radiation.

THE ANCIPITAL INTERSTELLAR MEDIUM
Adeo Varia Fortuna Belli Ancepsque Mars Fuit Ut Propius Periculum Fuerint Qui Vicerunt ... Livy, *Ab Urbe Condita*, Liber XXI

The contents of the previous two sections present a contradictory tale. The results of angular broadening measurements favor a quasi-Kolmogoroff spectrum, with an index in the range 3.7-3.9. Certain types of measurements on pulsar dynamic spectra, such as the spectral correlation function shape, support this result. On the other hand, the *very existence* of phenomena such as the drifting bands in dynamic spectra, and particularly quasi-periodic interference patterns such as those shown in Figure 8, seem to require the agency of steep spectra with $\alpha > 4$ unless our understanding of interstellar scintillation is completely defective. We may then refer to this condition, in which certain types of observations indicate a quasi-Kolmogoroff spectrum, while others suggest a steep spectrum, as indicative of an *ancipital* interstellar medium [1]. An extreme case of this ambiguous nature is to be found in observations of the pulsar 1133+16. In their Figure 8, Armstrong and Rickett[8] show that a quasi-Kolmogoroff spectrum with $\alpha = 3.6$ is a decidedly better fit to the spectral autocorrelation function than the Gaussian spectrum. However, a number of observers have documented the frequent presence of drifting bands and quasi-periodic interference patterns in the dynamic spectrum of this object, so strong evidence for a steep density spectrum along this line of sight is also present.

A possible clue to understanding the ancipital quality of interstellar scintillation is provided by the disparity in the strength of scattering of, on the one hand, the broadened sources discussed in Section 3, and on the other the pulsars displaying structured dynamic spectra. The product $C_N^2 z_{kpc}$ is in excess of 0.1 for the objects displaying angular broadening, whereas it is 10^{-4} for the pulsar 1133+16. We could then contend that an effective quasi-Kolmogoroff spectrum characterizes highly scattered lines of sight, whereas short lines of sight sample a medium whose spectrum is considerably steeper. A potential difficulty with this suggestion is the observation of interference patterns in the dynamic spectrum of pulsar 1933+16 (J.M. Cordes, private communication). The scattering of this object is intermediate to PSR 1133+16 and radio source 2013+370, and would seem to indicate that refractive effects do not dissapear with increasing strength of scattering .

There are at least two sensible explanations for the ancipital quality of the scintillations. First, it is possible that the two types of measurements refer to different regions of the interstellar medium. As discussed by Cordes,Weisberg, and Boriakoff[3], pulsar measurements indicate that turbulence in the interstellar medium is located in two types of regions, a diffuse distributed component termed "Type A" by those authors, and intense, small scale clumps of turbulence referred to as "Type B" turbulence. Cordes *et al* argue that highly scattered lines of sight are dominated by the Type B component. It is therefore possible that the Type B turbulence, probed by the angular broadening measurements, has a quasi-Kolmogoroff spectrum, while the diffuse Type A medium has fluctuations with a steeper power law, a quasi-Kolmogoroff spectrum enhanced at low wavenumbers, or a quasi-Kolmogoroff spectrum with an inner scale near the Fresnel scale[2].

A second and in many ways more exciting possibility is that we are seeing the effect of an outer scale to the turbulence. Refractive effects will be quenched if the multipath scale, $\theta_d z$ becomes comparable to or larger than the outer scale of the turbulence[10]. Gwinn *et al*[7] discuss this possibility in interpreting their data, and obtain an outer scale of $\sim 2 \times 10^{13}$ if this explanation is valid.

Such a scale is orders of magnitude smaller than those on which shear turbulence would be produced due to HII regions, differential galactic rotation, etc. A prevailing notion in this field, that the scintillation turbulence is the extreme high wave number end of an inertial subrange beginning on scales of hundreds of parsecs, would consequently be subject to extreme and sceptical scrutiny. Such an outer scale would, however, be quite consistent with a model in which density fluctuations are associated

---

[1]The english word ancipital is derived from the Latin *anceps*, meaning "two faced", and was used, for example, by Livy in his epitome of the Second Punic War.

with magnetohydrodynamic waves[19,20].

THE GALACTIC DISTRIBUTION OF TURBULENCE

Observations of diffractive interstellar scintillation also provide information on the distribution of turbulence in the galaxy. Work by several *comitatus* of investigators is progressing toward a "scattering map of the galaxy", analogous to that provided by HI surveys, etc. Such an investigation may well contribute at least as much to a knowledge of the genesis of the turbulence as precise measurement of the density spectrum. If we can identify regions of enhanced turbulence with specific astronomical objects, it will be much easier to divine the mechanism which generates the irregularities.

This section will be brief, since the topic is discussed in other presentations at this meeting. Perhaps the primaryresult of research to date is that the interstellar turbulence is highly inhomogeneous in its galactic distribution. This was mentioned by Rickett[21], and research in the subsequent decade has enforced this notion. Observations by Dennison *et al*[22] showed substantial changes in angular broadening on scales of only a few degrees. Recent observations revealing marked changes in angular broadening are reported at this meeting by Fey *et al*[23]. Dennison *et al* interpreted the variability as due to the interposition of intense, localized regions of turbulence. Cordes *et al*$^\beta$ reached the same conclusion through analysis of a large body of pulsar scintillation measurements.

Observations to date have yielded no positive indications as to the nature of these clumps. The vicinities of supernova remnants have been specifically examined for evidence of enhanced scattering due to magnetohydrodynamic wave activity or serendipitous mechanisms, but so far with negative results[19,24].

CONCLUSIONS AND EXHORTATIONS

The following seems to be a reasonable summary of the current status of the observations.

- Angular broadening measurements on highly scattered lines of sight favor quasi-Kolmogoroff spectra with indices roughly in the range 3.7-3.9.

- Certain facets of pulsar dynamic spectra also suggest quasi-Kolmogoroff spectra. We refer to the scaling of decorrelation bandwidth with frequency, the shapes of spectral autocorrelation functions, and perhaps even the fact that observed parameters, made from finite temporal averages, are even recognizably similar to theoretical entities employing the ideal of "ensemble average".

- Pulsar dynamic spectra also reveal phenomena, such as drifting frequency bands and quasi-periodic interference patterns, which *suis ipsis generibus* require steep density spectra.

- One may readily construct suggestions for resolving this ancipital situation, but it is unclear at present which, if any, of the possibilities are true.

- The spatial distribution of turbulent intensity (rigorously, the parameter C_N^2) is highly inhomogeneous, with the presence of both intense, localized clumps and a diffuse, relatively uniform background.

In conclusion, I would like to take an opportunity to encourage observational investigation of two matters which I feel would greatly help the progress of our understanding of interstellar turbulence.

First, it would be interesting to know how many pulsars display patently refractive dynamic spectra, and what fraction of the time they do so. At present it is clear that the quality of the spectra changes with observing season, and consequently it may be true that most or all low dispersion pulsars show bands or interference patterns some of the time. Cordes and Wolszczan[13] contend that this is the case. The kind of project proposed would help us to determine if refractive phenomena are episodic or represent the customary situation in scintillations. Such knowledge, in turn, would help us choose between the possibilities for steep spectra described by Coles *et al*[2].

Second, it would be desirable to see if we can obtain evidence for an outer scale to the turbulence along the line of sight to a pulsar. Such an endeavor is encouraged by oblique evidence for such a scale, some of which was mentioned above. Probably the best way to approach this problem would be to acquire simultaneous angular broadening measurements, to set the multipath scale, and dynamic spectral observations. A clear and persistent absence of refractive spectral features in a pulsar with a large multipath scale could be interpreted as evidence for an outer scale. As discussed above, this would be of great import for mechanisms of turbulence generation.

ACKNOWLEDGEMENTS

The author acknowledges support from the National Aeronautics and Space Administration through grants NAGW-806 and 831. Figures 3, 5, and 6 were from unpublished data of J.M. Cordes, J.M. Weisberg, and V. Boriakoff associated with reference (3).

REFERENCES

1. D. Montgomery, M.R. Brown, and W.H. Matthaeus, J. Geophys. Res., 92, 282 (1987).

2. W.A. Coles, R.G. Frehlich, B.J. Rickett, and J.L. Codona, Astrophys. J., 315, 666 (1987).

3. J.M. Cordes, J.M. Weisberg, and V. Boriakoff, Astrophys. J., 288, 221 (1985).

4. B.J. Uscinski, The Elements of Wave Propagation in Random Media (McGraw Hill, N.Y., 1977), p55.

5. S.R. Spangler and J.M. Cordes, Astrophys. J., (in press) (1988).

6. P.N. Wilkinson, R.E. Spencer, and R.F. Nelson, in IAU Symposium 129, *The Impact of VLBI on Astrophysics and Geophysics* (Dordrecht:Reidel), 1988.

7. C.R. Gwinn, J.M. Moran, M.J. Reid, M.H. Schnepps, R. Genzel, and D. Downes, in IAU Symposium 129, *The Impact of VLBI on Astrophysics and Geophysics* (Dordrecht:Reidel), 1988.

8. J.W. Armstrong and B.J. Rickett, Mon. Not. R. Astro. Soc., 194, 623 (1981).

9. A. Wolszczan, Mon. Not. R. Astro. Soc., 204, 591 (1983).

10. J.M. Cordes, A. Pidwerbetsky, and R.V.E. Lovelace, Astrophys. J., 310, 737 (1986).

11. A. Hewish, Mon. Not. R. Astro. Soc., 192, 799 (1980).

12. V.I. Shishov, Astr. Zh., 50, 941 (1974).

13. J.M. Cordes and A. Wolszczan, Astrophys. J. (Lett.), 307, L27 (1986).

14. J.A. Roberts and J.G. Ables, Mon. Not. R. Astro. Soc., 201, 1119 (1982).

15. A. Hewish, A. Wolszczan, and D.A. Graham, Mon. Not. R. Astro. Soc., 213, 167 (1985).

16. R.W. Romani, R. Narayan, and R. Blandford, Mon. Not. R. Astro. Soc., 220, 19 (1986).

17. F.G. Smith and N.C. Wright, Mon. Not. R. Astro. Soc., 214, 97 (1985).

18. J. Goodman and R. Narayan, Mon. Not. R. Astro. Soc., 214, 519 (1985).

19. S.R. Spangler, A.L. Fey, and J.M. Cordes, Astrophys. J., 322, 909 (1987).

20. S.R. Spangler, these proceedings, 1988.

21. B.J. Rickett, Ann. Rev. Astr. Ap., 15, 479 (1977).

22. B. Dennison *et al* Astrophys. J., 313, 141 (1987).

23. A.L. Fey, S.R. Spangler, and R.L. Mutel, these proceedings, 1988.

24. S.R. Spangler, R.L. Mutel, J.M. Benson, and J.M. Cordes, Astrophys. J., 301, 312 (1986).

II. PHYSICS OF SCATTERING MEDIA

Cosmic Rays and the Physics of Interstellar Turbulence

J. R. Jokipii
University of Arizona

Abstract.
The transport of cosmic rays in the interstellar medium is reviewed with emphasis on interactions with the turbulent interstellar magnetic field. It is shown that the standard picture of cosmic-ray transport suggests strongly the existence of a smooth turbulence spectrum over the range of scales between 10^{12} and 10^{19} cm. This, coupled with observations of radio wave scattering and other direct measurements suggests a smooth, power-law turbulence spectrum over the range of scales from 10^9 to 10^{19} cm, with the index of the power law being close to that of the Kolomogorov equilibrium subrange. Attributing this to a turbulent cascade leads to severe difficulties. Suggestions for avoiding these difficulties are discussed.

1. Introduction.

The structure and dynamics of interstellar turbulence is a problem which is not only of interest in its own right, but one which provides important insights for other areas of inquiry ranging from plasma physics to the structure of the galaxy. Observations of the interstellar scattering of radio waves yield information about density fluctuations on scales of the order of $10^9 - 10^{14}$ cm. If one approximates the spectrum of density fluctuations as a power law in spatial wavenumber magnitude k ,

$$P(k)dk = A_n^2 k^{-\alpha} dk, \qquad (1)$$

then the radio observations suggest that[1] α is less than 2. The data are consis-

---

[1] Note that some of the scintillation literature considers the three-dimensional spectrum, $P_{3N}(k)d^3k = C_N^2 k^{-\beta}d^3k$ (e.g. Narayan, this volume); thus, $A_n^2 = 4\pi C_N^2, \alpha = \beta - 2$.

tent with $\alpha = 5/3$, the Kolmogorov exponent, over the range of wavenumbers covered. Information concerning much larger scales $\approx 10^{18}$ cm is provided by direct observations of interstellar gas velocities, magnetic field, etc. These data indicate a level of fluctuations at scales of the order of pc which is surprisingly close to the extrapolation of the scintillation data to these scales, using the Kolmogorov exponent (Lee and Jokipii, 1976, Armstrong, Cordes and Ricket, 1981). One is then naturally led to suggest a single power-law spectrum over the entire range of some 10 decades in wavenumber from 10^9 cm to 10^{19} cm, although observations are not currently available for the range of 10^{14} to 10^{18} cm. It does not appear possible any time soon to fill in this "gap" in the observations using radio data. In this paper observations of the cosmic-ray energy spectrum, together with some general considerations of charged-particle transport are shown to provide some support for a continuous power-law spectrum across the gap. The theoretical difficulties associated with maintaining such a spectrum by a turbulent cascade will then be summarized, together with recent suggestions for avoiding them.

2. Cosmic-Ray Confinement and Transport in the Galaxy

a. Observations

Cosmic rays are energetic charged particles which originate in the galaxy. They consist mainly of protons, with the nuclear composition being essentially that of normal matter, with a modest enhancement of low-ionization-potential species. The energy spectrum of the particles is shown in figure 1. The part of the spectrum corresponding to energies between 10^8 and 10^{15} eV can be termed the "normal" galactic cosmic rays. They are thought to originate in supernova explosions in the galaxy, and arrive at the Earth after propagating for more than 10 million years in the interstellar medium.

I will restrict my attention to those particles with energies above about 10^9 eV to avoid the part of the spectrum which is seriously distorted by the solar wind. These "unmodulated" galactic cosmic rays are observed to be highly isotropic,($\delta < 10^{-3}$ at $\approx 10^{12}$ ev). Furthermore, evidence from observed γ rays and synchrotron emission indicates that they are uniformly distributed throughout the galactic disk. Finally, observations of unstable isotopes in meteorites show that these cosmic-rays have been constant to within about 50 %, at the solar system, for approximately the last 10^9 years. The energy spectrum in interstellar space is apparently very smooth, with no significant departure from the power law $T^{-2.6}$ in the energy range corresponding to scales of 10^{12} to 10^{19} cm.

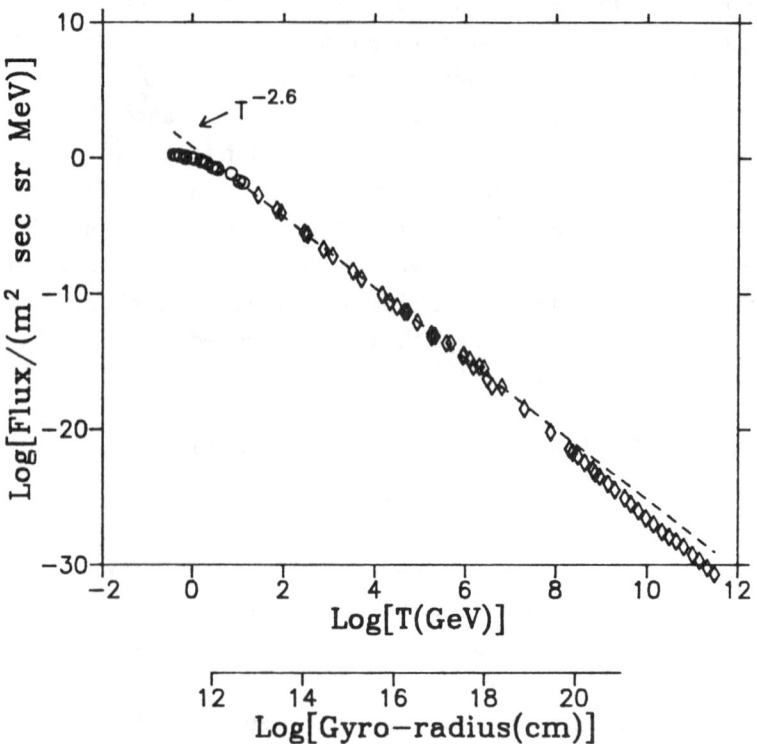

Figure 1. The galactic cosmic-ray spectrum for energies greater than 500 Mev. The dashed line is a power law with index -2.6. The second abscissa illustrates the gyro-radius of a proton with the given energy in a nominal 3×10^{-6} Gauss interstellar magnetic field. Note the change in slope at about 10^{15} eV. The turn over of the spectrum at low energies is due to solar effects on the intensity observed at Earth. Data compiled from Meyer (1969) and Linsley (1980).

Next consider the observed ratio of secondary-to-primary cosmic-ray nuclei as a function of energy per nucleon. Secondaries are produced by spallation of heavier cosmic-ray nuclei by collisions with the ambient interstellar gas particles. Observations show that above roughly 5 GeV per nucleon, the ratio of secondaries to primaries decreases monotonically with increasing energy, but below $1-2$ GeV per nucleon this ratio is approximately independent of energy. It is useful to consider the light (L) nuclei, which are not present in the sources. A fit to the data presented by Fontes, Meyer, and Perron (1978) yields for the L to $C+O$ ratio.

$$\frac{n_L}{n_{C+O}} \approx T^{-0.4} \qquad (2)$$

above 2 GeV per nucleon, with an error of about ±0.1 in the exponent. This is generally interpreted as a result of an energy-dependent leakage as discussed below.

b. Cosmic-Ray Transport and Confinement

Since the seminal work of Fermi (1949) it has been realized that the transport of cosmic rays in the interstellar medium is determined by the irregular interstellar magnetic field. Hence one may use cosmic rays as a probe of interstellar turbulence. The information so obtained is complementary to that obtained using radio astronomy techniques and, as we will see, provides information concerning the "gap" between scales of 10^{14} and 10^{18} cm above.

The theory of cosmic-ray transport is quite well developed. Under conditions such as those expected in the interstellar gas, the particles are scattered and made nearly isotropic by irregularities in the ambient magnetic field. The resulting transport is diffusive and can be described in terms of a diffusion coefficient κ, (which is actually a tensor) together with effects associated with convection by the ambient fluid. Under quite general circumstances one finds that the diffusion at any given particle energy is determined by magnetic irregularities having scales comparable to the gyro-radius of the particles in the average magnetic field. The lower scale in figure 1 shows the gyroradii of cosmic-ray protons as a function of energy in a nominal 3×10^{-6} Gauss interstellar magnetic field. It is apparent that the observed cosmic rays can in principle provide information concerning turbulent fluctuations in the magnetic field at scales ranging from 10^{12} to 10^{19} cm, which spans the gap in the radio scattering data discussed above.

Although it is possible to be more sophisticated, it is adequate for present purposes to describe the propagation of these cosmic rays and their confinement to the galaxy in terms of a relatively simple model. In this model, the galaxy is taken to be a leaky box containing cosmic rays of species i with a density $n_i(T)$. The loss of these particles is described by a mean leakage time τ (which could be related to a diffusion coefficient κ by $\tau \approx L^2/\kappa$, where L is a characteristic scale of the galactic confinement region). If there is a source of particles $Q_i(T)$, conservation of particles is described by the equation

$$\frac{\partial n_i}{\partial t} = -\frac{n_i(T)}{\tau} + Q_i(T). \tag{3}$$

This will be adequate for our needs.

Observations of ^{10}Be (e.g. Garcia-Munoz, Mason and Simpson, 1975) show that the mean trapping time τ for cosmic rays in the galaxy is more than 10^7 years, so the present distribution of cosmic rays reflects a steady state between sources and losses, and we have the simple result from equation (2).

$$\begin{aligned} n_i(T) &= \tau(T) Q_i(T) \\ &\approx \frac{L^2}{\kappa(T)} Q_i(T). \end{aligned} \tag{4}$$

The equilibrium energy spectrum depends on both the source spectrum and the energy dependence of the leakage time. Thus, for example if $\tau \propto T^{-\ell}$ (we expect $\ell > 0$ since higher-energy particles escape more quickly) and if $Q_i \propto T^{-\gamma}$, we have $n_i \propto T^{-(\gamma+\ell)}$. The spectrum is steepened by the loss process.

Because the production cross sections at energies $\gtrsim 2\ GeV$ are roughly independent of energy, the dependence given in equation (4) would be produced if the leakage time were approximately proportional to $T^{-0.4}$ above $\sim 1-2\ GeV$. A number of authors have carried out similar analyses, generally concluding that the leakage time above $\sim 1-2\ GeV$ scales as T^{-s}, where s ranges between 0.4 and 0.6 (e.g., Ormes and Freier 1978; Caldwell and Meyer 1977; Audouze and Cesarsky 1973). The observed energy spectra of other nuclei and electrons also fall quite reasonably into this overall picture (see, e.g., Jokipii and Higdon, 1979).

c. Acceleration of Cosmic Rays.

Galactic cosmic rays are currently believed to be accelerated by collisionless shock waves (Axford, Leer, and Skadron 1977; Bell 1978a, b; Blandford and

Ostriker 1978). Subject only to a few quite reasonable restrictions, which essentially amount to assuming a plane shock and validity of the diffusion approximation in cosmic ray scattering by magnetic irregularities, one finds for the spectrum of accelerated particles above an injection energy T_i,

$$Q_{sh}(T) \approx A(T + m_0 c^2)(T^2 + 2m_0 T c^2)^{-(2r+1)/(2r-2)} \qquad (5)$$

where $r = u^-/u^+ = \rho^+/\rho^-$ is the compression ratio across the shock, A is a constant, and m_0 is the rest mass. r approaches 4 for strong shocks, and blast waves from supernovae or other efficient shocks are expected to be relatively strong. Hence, one expects, quite generally, that the accelerated energy spectrum for $T \tilde{\gg} 1\ GeV$ per nucleon to be

$$Q(T) \approx A T^{(-2+\epsilon)}, \qquad (6)$$

where $\epsilon = (4-r)/(r-1)$ is presumably small compared with unity. This shape is independent of the particle propagation parameters, insofar as the basic assumptions are satisfied. Quite naturally, then, from equation (4) we would expect the observed $\sim T^{-2.6}$ primary nucleon energy spectrum from a source spectrum of the form given in equation (6) with $\epsilon \approx 0.2 (r \approx 3.5)$, and $\ell \approx .4$, which is not unreasonable.

We note that although this source spectrum is located at the shock fronts, over the $\gtrsim 10^7$ year lifetime of a typical cosmic-ray particle, one expects that much of the interstellar medium will be traversed by shocks. Hence, it is reasonable to assume a smooth source distribution in the disk.

These considerations lead to the quite general conclusion that the shock-wave acceleration model with strong shocks ($r \approx 3.5$), coupled with the energy-dependent loss consistent with that derived from the secondary-to-primary ratio, provides a reasonable interpretation of the general energy spectra of cosmic-ray nuclei and electrons at energies greater than a few GeV. There are areas of disagreement between the observations and this simple picture, but can be argued that they do not contradict the general picture describe here.

3. Consequences for the Spectrum of Interstellar Turbulence.

The above discussion concerning the physics of cosmic-ray transport in the interstellar medium has implications for interstellar turbulence. First the mean leakage time is several thousand rectilinear transit times across the galaxy. This fact together with the observed high degree of isotropy of the radiation implies scattering by turbulent fluctuations in the magnetic field. The particles propagate in the galaxy and escape by diffusion. I showed above that the loss time for particles is given by L^2/κ. We have also seen that the secondary to primary ratio implies that the loss time τ_L is proportional to $T^{-0.4}$. Combining these two expressions we then say that the diffusion coefficient

$$\kappa \propto T^{0.4}, \tag{7}$$

again with an uncertainty of about 0.1 in the exponent. The relationship of the cosmic-ray diffusion coefficient to the power spectrum of magnetic field fluctuations depends on the nature of the magnetic turbulence. But the general nature of the conclusions are relatively insensitive to the details.

If we use quasilinear theory and consider the particle orbit in a magnetic field which varies only along the average magnetic field direction (slab model), we come up with the result that the rate of scattering in pitch angle θ (relative to the average magnetic fields) of relativistic particles having momentum p is given by (Jokipii, 1971, equation (61))

$$\frac{1}{\tau_{scat}} \sim \frac{<\Delta\theta^2>}{\Delta t} \propto \frac{1}{p^2} P_B(k = \frac{1}{r_c cos(\theta)}). \tag{8}$$

This then implies that at relativistic energies, the diffusion coefficient κ can be related to the particle energy and the power-law exponent for the turbulence α defined above through

$$\kappa \propto T^{2-\alpha}. \tag{9}$$

Now, if we were to assume that the magnetic spectrum in interstellar space were proportional to $k^{-5/3}$, as is the case for Kolmogorov turbulence, we would then expect that κ is proportional to $T^{0.33}$. It is of interest that since this is in the range of values expected from the above analysis of the secondary to primary ratio, the picture of cosmic-ray confinement by scattering by a Kolmogorov spectrum of magnetic irregularities is attractive. This must be tempered by the rather restricted range of circumstances where quasi-linear theory is a good approximation. However, in this regard, one may draw a somewhat more general conclusion. Independent of the detailed applicability of quasilinear theory, one expects the cosmic rays to be most affected by irregularities with scales near their cyclotron radius. The observational fact that the spectrum as shown in figure 1 is really quite smooth as a function of energy indicates that the spectrum of magnetic irregularities is also smooth.

On similar, general grounds one may argue that the spatial dependence of the irregularity spectrum cannot be too "clumpy", for this would lead to a clumpy cosmic-ray distribution, which is contrary to observations. Since cosmic rays come from no more than about 1000 pc (Jokipii and Parker, 1968), the clumps must be substantially closer together than this distance.

The next question we can ask, about the implications of cosmic- rays for the interstellar turbulence, is what do these conclusions concerning the magnetic spectrum imply about **density** fluctuations in the interstellar gas? Direct observations of the solar wind show that there the magnetic field and density both have power-law spectra with nearly the same slopes (Goldstein and Siscoe, 1971). Analogy would suggest that both the interstellar magnetic field and density exhibit the same power law spectrum. There are also theoretical arguments which suggest that the spectrum of density fluctuations will have a shape similar to that of velocity in the magnetic field. Higdon (1984), has considered the convection of a passive additive and has shown that this gives a Kolmogorov spectrum. He then argues that the density fluctuations can be regarded as a passive additive, if there are entropy fluctuations in the interstellar gas. Finally, Montgomery, Brown and Mattheaus (1987), using a magnetohydrodynamic theory of a fluid have shown that a Kolmogorov spectrum should give rise to a similar shape in the fluctuations in density. So we conclude then that the cosmic ray observations also point to the existence of a smooth spectrum of density fluctuations in the interstellar medium over scales of 10^{12} to 10^{18} cm.

These arguments lead to the following four conclusions concerning the turbulence in the interstellar medium.

[a] Application of quasilinear theory to cosmic-ray transport suggests a general background level of magnetic fluctuations and is consistent with a Kolmogorov type spectrum.

[b] More generally, the observed smoothness of the cosmic-ray energy spectrum over the range of energies from 10^9 to 10^{15} ev, points strongly to the

existence of a continuous spectrum of magnetic fluctuations in the interstellar medium, from scales of 10^{12} cm. to 10^{19} cm. This range of scales spans the observational gap in observations of interstellar turbulence from radio scintillation observations.

[c] The observed spatial homogeneity of cosmic rays in the galaxy leads to the conclusion that the turbulence in the magnetic field cannot be too "clumpy". Otherwise there would be significant large spatial variations in the cosmic-rays.

[d] One expects the spectrum of density fluctuations to be similar in shape to that of the magnetic field.

The cosmic-ray data, then, taken together with the radio observations, suggest the existence of a smooth irregularity spectrum in the interstellar gas, ranging over some ten decades in wave number from scales 10^{18} cm. down to perhaps 10^8 cm.

4. Physics of Interstellar Turbulence.

In this section I discuss, briefly, the physical consequences of assuming an overall turbulence spectrum in the galaxy, as suggested in the previous section. Although the picture proposed above is extremely suggestive and appears to be consistent with a variety of quite different kinds of observations, there are severe physical problems with actually producing such a spectrum. The only plausible method which has been suggested for producing a smooth, power-law irregularity spectrum over many decades in wave number is through a cascade from larger to smaller scales. For if one has a variety sources, producing irregularities at different scales, there is difficulty in understanding why the spectrum is smooth. The problem with a cascade lies not so much in generating the turbulence at the larger scales, but in the fact that there are many processes which can act to absorb energy from the turbulence at intermediate scales, stopping the cascade.

A general, generation mechanism for large scale turbulence arises from super novae explosions and other large scale motions which produce velocity gradients and shear. These will have an extraordinarily high Reynolds number, and should therefore generate turbulence. This particular mechanism can easily supply the amount of energy required for a Kolmogorov cascade at the amplitude observed.

Consider now various damping mechanisms which have been discussed. In practice what happens is that because the interstellar gas is essentially collisionless, there exists a wide variety of methods for particles to remove energy from the turbulent fluctuations.

One such mechanism is the second order Fermi acceleration, which can be used to accelerate cosmic rays. Since the turbulent cascade will produce waves propagating in random directions, the conditions for second order Fermi acceleration are met (Jokipii, 1977). One may write an equation for the rate of energy gained in terms of the diffusion coefficient and the Alfvén speed V_A,

$$\frac{\dot{T}}{T} \simeq 1.7 \frac{V_A^2}{\kappa}. \tag{10}$$

If the power spectrum of the magnetic regularities has the following form

$$P_B(k) \simeq \frac{A}{(1 + k^2 L_c^2)^{5/6}}, \tag{11}$$

with $L_c \simeq 30$ pc and with A chosen so that $< B^2 >^{1/2} \simeq 3 \times 10^{-6} G$, we find that the characteristic time scale for acceleration of cosmic-rays of the order of GeV energy is approximately 10^6 years. This is much too rapid, and would absorb the energy from the turbulent cascade at scales of the order of 10^{13} cm.

Other mechanisms which have been discussed by a number of authors involve Landau damping and thermal conduction. I will not discuss these in detail here because they have been discussed adequately in the literature. But there is no doubt that if one applies these mechanisms to the waves associated with the turbulence, they effectively prevent a turbulent cascade to scales much smaller than 10^{13} cm. Recently Bykov and Toptygin (1987) have suggested that non-linear suppression of the Landau damping may occur, but even they have difficulty producing scales less than about 10^{13} cm. We can therefore conclude that a straightforward application of the physics of wave propagation in a collisionless gas suggests that the turbulent cascade hypothesis for producing the irregularity spectrum has serious difficulties.

Higdon (1984) suggested that all density fluctuations have wave vectors normal to the local magnetic field. Hence, these locally two dimensional fluctuations cannot effectively damp because of the suppression of particle motions normal to the magnetic field. Higdon's approach has proved difficult to quantify and the question of cosmic-ray transport in such a magnetic field has not been addressed. The work of Montgomery, Brown and Mattheaus (1985), which appears to be based on similar physics also looks promising.

Finally, at this meeting there was a presentation by Gibson (see also Gibson, 1987), concerning the possibility that what is observed is "fossil" turbulence. This is an analogy with processes occurring in laboratory fluids and in the oceans, where a region of active generation of turbulence has occurred, but the associated motions have damped away. This then leaves fluctuations in the density and temperature of the fluid (and, presumably in any coupled magnetic field), which exhibit the Kolmogorov spectrum, but which no longer actively damp or decay. Such a proposal would be very attractive for the interstellar medium, if a way could be found to implement it. For then, one could say that the fluctuations of the density and the magnetic field are simply relicti of a previous turbulent state, and the questions of the turbulent cascade and the problems with the waves can be pushed to a different arena, or perhaps done away with altogether. This is an interesting idea which deserves further attention.

5. Conclusions.

Consideration of cosmic ray observations and theories of their transport in the interstellar gas provides further evidence concerning the spectrum of irregularities in the interstellar gas. The cosmic-ray and radio-wave scattering data, taken together, suggest that there is a relatively homogeneous, smooth spectrum of magnetic irregularities from scales of the order of 10^9 cm to 10^{19} cm. The data are consistent with a power law in wavenumber magnitude with index $\alpha = 5/3$, which corresponds to the Kolmogorov spectrum.

The data suggest that the observed spectrum is the result of a turbulent cascade. However formidable theoretical difficulties exist in maintaining such a cascade against various collisionless damping mechanisms. Mechanisms for avoiding this difficulty include modifying the nature of the fluctuations to minimize damping or postulating that the fluctuations are "fossil" so that there is no longer any turbulent cascade occuring.

Acknowledgements.

This work is Contribution 88-06 of the University of Arizona Theoretical Astrophysics Program and was supported in part by the National Aeronautics and Space Administration under Grant NsG-7101 and by the National Science Foundation under Grant ATM-8618260.

REFERENCES

Armstrong, J. W., Cordes, J. M., and Rickett, B. J. 1981, *Nature*, **291**, 561.
Audouze, J., and Cesarsky, C.J. 1973, *Nature Phys. Sci.*, **241**, 98.
Axford, W.I., Leer, E., and Skadron, G. 1977, *Proc. 15th International Cosmic Ray Conference*, Plovdiv, Bulgaria, **2**, 273.
Bell, A.R. 1978a, *M.N.R.A.S.*, **182**, 147.
───── 1978b, *M.N.R.A.S.*, **182**, 443.
Blandford, R.D., and Ostriker, J.P. 1978, *Ap. J. (Letters)*, **221**, L29.
Bykov, A. M. and Toptggin, I. N. 1987, *Astroph. Sp. Sci.*, **138**, 341.
Caldwell, J.H., and Meyer, P. 1977, *Proc. 15th International Cosmic Ray Conference*, Plovdiv, Bulgaria, **1**, 243.
Cesarsky, L.J. 1980, *Ann. Rev. Astron. Ap.*, **18**, 289.
Fermi, E. 1949, *Phys. Rev*, **75**, 1169.
Fontes, P., Meyer, J.P., Peroon, C. 1977, *Proc 15th Internatinal Cosmic Ray Conference*, Plovdiv, Bulgaria, **2**, 234.
Garcia-Munoz, M., Mason, G. M., and Simpson, J. A. 1975 *Ap. J. (Letters)*, **201**, L142.
Goldstein, B., and Siscoe, G. L. 1972, in *Solar Wind* ed. by C. P. Sonett, P. J. Coleman and J. M. Wilcox, NASA SP-308, p. 506.
Gibson, Carl H. 1987, *J. Geophys. Res.*, **92**, 5383.
Higdon, J.G. 1984, *Ap. J.*, **285**, 109.
───── 1986, *Ap. J.*, **309**, 342.
Jokipii, J.R. and Higdon, J.C. 1979, *Ap. J.*, **228**, 293.
Jokipii, J.R. 1971, *Rev. Geophys. and Sp. Phys.*, **9**, 27.
Jokipii, J.R. and Parker, E.N. 1969, *Ap. J.*, **155**, 799.
Jokipii, J.R. 1977, *Proc 15th Internatinal Cosmic Ray Conference*, Plovdiv, Bulgaria, **2**, 109.
Lee, L.C. and Jokipii, J.R. 1976, *Ap. J.*, **206**, 735.
Linsley, J. 1980, *IAU Symposium 94, Origin of Cosmic Rays*, ed. G. Setti, G. Spada, and R. A. Wolfendale (Dordrecht, Reidel) p. 53.
Meyer, Peter 1969, *Ann. Rev. Astron. Astroph.*, **7**, 1.
Montgomery, D., Brown, M. R., and Mattheaus, W. H. 1987, **J. Geophys. Res.**, **92**, 282.
Ormes, j.R. and Freier, P. 1978, *Ap. J.*, **222**, 471.

TURBULENT MAGNETOHYDRODYNAMIC DENSITY FLUCTUATIONS
David Montgomery, Dartmouth College

MHD turbulence theory has developed mostly be generalizing Navier-Stokes results, almost always incompressible ones. It has recently been possible to develop[1] a slightly-incompressible theory of MHD density fluctuations by what is essentially a generalization of Lighthill's method. An approximately incompressible MHD turbulence field drives a parasitic density field, the fluctuation spectrum of which can be expressed in terms of kinetic and magnetic spectra. If the incompressible MHD spectrum is Kolmogoroff-like, the inertial-range results can be summarized by saying that the Fourier-transformed density fluctuation, $\delta\rho_{\mathbf{k}}$, is proportional to $-(B^2)_{\mathbf{k}}$, the kth Fourier component of the <u>square</u> of the variable magnetic field.[2] Making the quasi-normal approximation on expectations of products of four $\mathbf{v}$ and $\mathbf{B}$ field Fourier coefficients, a $k^{-5/3}$ omni-directional ($k^{-11/3}$ modal) density fluctuation spectrum. Even without enough spatial scale separation for the Kolmogoroff assumptions to apply, it is still possible to demonstrate by numerical solution[2] of the MHD equations that the connection between $\delta\rho_{\mathbf{k}}$ and $(B^2)_{\mathbf{k}}$ is valid for low enough Mach number and high enough beta (ratio of mechanical to magnetic pressure).

The most difficult assumption to justify, for the magnetohydrodynamics of the interstellar medium, is the use of an equation of state, $p_{mechanical} = p(\rho)$, which uniquely connects the density and mechanical pressure. Even assuming the existence of an equation of state, it will in general have to be of the form $p=p(\rho,s)$, where s is the specific entropy. The assumption is thus basically one of isentropic (or isothermal) MHD flow. At high Reynolds numbers, the entropy is produced mainly in the dissipation range, due to the action of thermal conductivity and resistivity. If the inertial-range hydrodynamic time scales are faster than the times necessary for the entropy to travel back up to the inertial length scales, the approximation would apparently be justified. It would also be justified (via an isothermal equation of state) if the inequality were sharply reversed; but not unless one of the two inequalities were satisfied would any equation of state be plausible. Until we know more about the thermodynamic parameters of the interstellar medium, this will remain an open question.

[1] D. Montgomery, M.R. Brown, and W.H. Matthaeus, J. Geophys. Res. <u>92</u>, 282 (1987).

[2] J.V. Shebalin and D. Montgomery, "Turbulent Magnetohydrodynamic Density Fluctuations", to be published in J. Plasma Phys., 1988 (in press).

INTERSTELLAR ELECTRON DENSITY FLUCTUATIONS DUE TO COSMIC-RAY ACCELERATION AT SUPERNOVA REMNANT SHOCK WAVES

Claire E. Max, Andrew Zachary[*], and Jonathan Arons[+]
Institute of Geophysics and Planetary Physics
Lawrence Livermore National Laboratory, Livermore, CA 94550

ABSTRACT

We have performed computer simulations to investigate whether sizeable electron density fluctuations might be produced in the process of cosmic-ray acceleration at supernova remnant shock waves. The hypothesis is the following: Cosmic-ray acceleration via a Fermi I mechanism leads to large-amplitude Alfven waves upstream of a supernova remnant shock wave. If the Alfven waves reach a large enough amplitude, they can drive sound waves in the interstellar medium. The fluctuations in the electron density due to these sound waves will then contribute to the observed interstellar scintillation. Our simulations suggest that this mechanism may be a plausible one. Issues remaining to be addressed include the resulting filling factor, and the integrated strength C_n^2 to be expected for the turbulence.

INTRODUCTION

Radio observations of interstellar scintillation[1] suggest the existence of localized regions having particularly intense fluctuations in the electron density. These regions of intense electron-density turbulence might be near HII regions, supernova remnants, or other point-sources of large energy input into the interstellar medium. Evidence suggests[1] an association with Population I objects, with scale height ≤ 100 pc and filling factor of $10^{-4} - 10^{-2}$. In the present paper we investigate one mechanism by which turbulent electron density fluctuations might be produced: during cosmic-ray acceleration in supernova remnant shock waves. If effective, this mechanism would lead to large electron-density turbulence in a spherical shell surrounding expanding supernova remnants. The sufficiency of this explanation for the entire clumped component of C_n^2 would then rest upon the probable geometrical filling-factor of the interstellar medium with the resulting turbulence, and its expected integrated strength.

In this short contribution, we consider the mechanism by which electron density fluctuations would be generated near supernova remnant shocks, and give preliminary results on its spectrum. Calculations of the geometrical filling factor and of the expected integrated turbulence strength are left for a future publication.

[*] Present address: Dept. of Astronomy, University of Chicago, Chicago, IL 60637
[+] Also: Depts. of Astronomy and Physics, University of California, Berkeley, CA 94720

EXCITATION OF ALFVEN WAVES BY STREAMING COSMIC-RAYS

We consider the region upstream of a supernova shock wave, in which cosmic-ray acceleration by a Fermi Type I mechanism is assumed to be taking place[2,3]. In the upstream region we postulate a self-consistent distribution consisting of anisotropic cosmic-rays and of the Alfven-waves which they excite. Our computational model[4] is a hybrid one, consisting of an MHD fluid background representing the interstellar medium, and of a system of 32,000 particle-in-cell discrete particles representing the cosmic-ray ions. The cyclotron orbits of the cosmic-rays are followed with high accuracy, and an averaging is performed over their orbits to calculate the self-consistent currents and magnetic fields. The computation is one-dimensional in space, but three-dimensional in velocity. We solve an initial-value problem[5] in which the magnetic field is uniform at t=0, and the cosmic-rays have a power-law distribution f(p) proportional to p^{-4} from 100 MeV/c to 2 GeV/c; p is the magnitude of the momentum. The initial cosmic-rays are isotropic in a frame drifting relative to the background interstellar medium at a speed 10 times the Alfven speed, and have a density $n_{cr} = 4 \times 10^{-3} n_{ism}$. Magnetic and thermal pressures of the background interstellar medium are equal; the Alfven speed is $v_A = 10^{-2} c$.

Figure 1 shows the time evolution of unstable Alfven waves in the fastest-growing mode, and of the sound waves which accompany them. The horizontal axis is the time measured in units of the interstellar ion cyclotron period. The sound waves grow up somewhat later in time than the Alfven waves, presumably because they are driven either by the resonant or non-resonant coupling of Alfven waves. The growing Alfven waves scatter the cosmic-rays in pitch-angle, making their distribution function more isotropic with time.

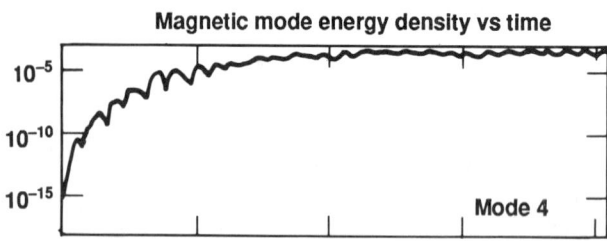

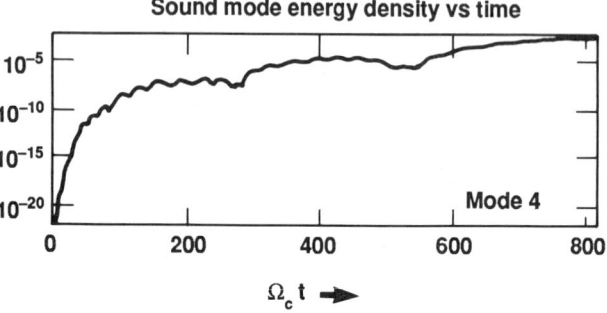

Figure 1
Temporal evolution of Alfven and acoustic waves

Figure 2 shows the k-spectrum of the sound waves. They peak at wavenumbers comparable to the inverse gyroradius of the lowest-energy cosmic rays, denoted on the Figure by r_L. For our lower cut-off of 100 GeV/c for the cosmic-ray distribution function, this implies wavelengths on the order of 10^{11} cm for the sound waves and their accompanying electron density fluctuations. Physically, this is because the Alfven waves are driven by a gyro-resonant instability of the cosmic-ray distribution function.

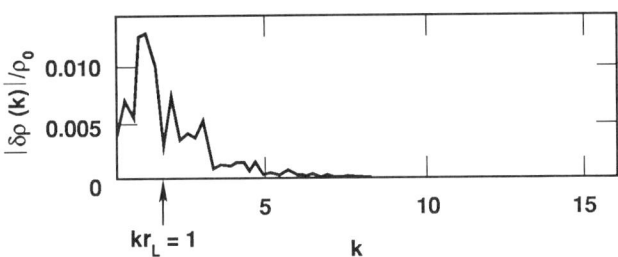

Figure 2
Acoustic wave spectrum

Such wave-lengths are short, and correspond to energy input on the high-wavenumber end of the spectrum of electron density turbulence in the interstellar medium. Thus if cosmic-ray acceleration at supernova remnant shocks is responsible for significant energy input into the spectrum of interstellar turbulence, this energy input would be on the large-k end of the spectrum. Such a notion is in contrast to the more conventional Kolmogorov picture, in which the largest scale eddies provide the energy input, which is later coupled into smaller and smaller spatial scales.

Figure 3 illustrates the space-time evolution of the magnetic field (left panel) and the interstellar medium density (right panel), for a computational example having $n_{cr} = 0.04\ n_{ism}$. At early times the magnetic field grows up according to a linear instability; the sound-wave amplitude follows somewhat later. One sees sound waves and Alfven waves propagating in the direction of the initial cosmic-ray anisotropy, which we call the "forward" direction. However in this example the Alfven waves reach large enough amplitude to also couple to backward-propagating waves, which one can see faintly as crests moving from the lower right toward the upper left in Figure 3.

The amplitude of the density fluctuations produced during cosmic-ray acceleration can be substantial. For a cosmic-ray density $n_{cr} = 4 \times 10^{-3}\ n_{ism}$, and other computational parameters as quoted above, the rms and peak values of the density fluctuation amplitudes were 4% and 10% respectively, relative to the density of the background interstellar medium. For a second computational example having ten times larger density ($n_{cr} = 0.04\ n_{ism}$), the rms and peak density fluctuation amplitudes were 20% and 50% respectively, relative to the density of the background interstellar medium.

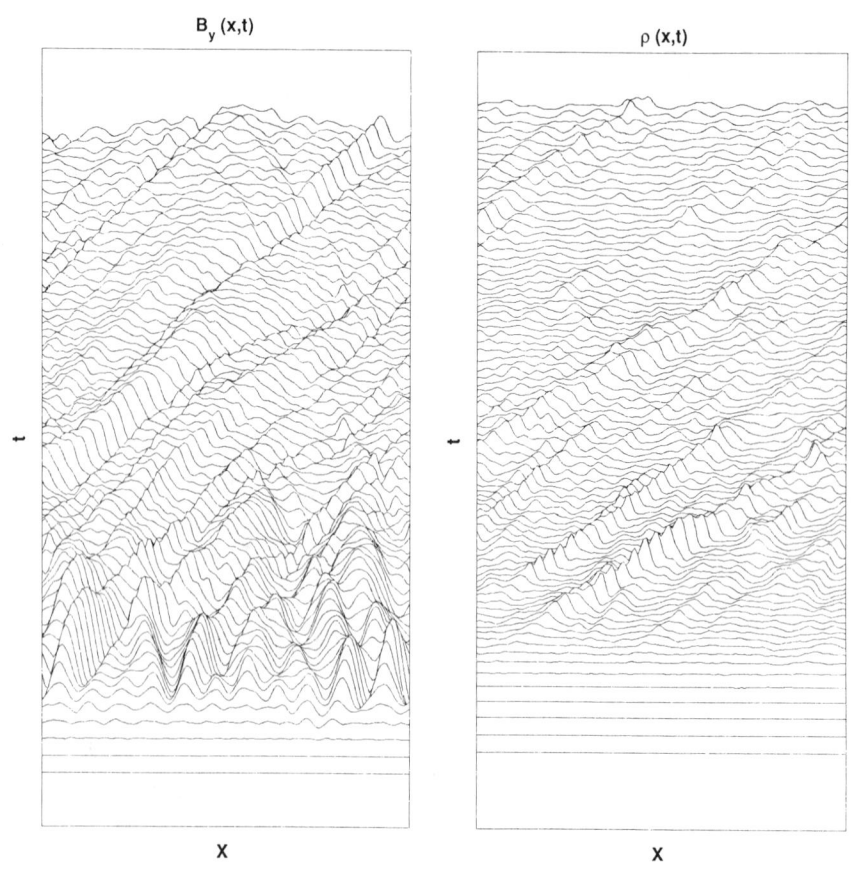

Figure 3
Space-time evolution of magnetic field (left)
and interstellar medium density (right)

Radio scintillation data[1] suggest rms amplitudes
$\langle \delta n_e^2 \rangle^{\frac{1}{2}} = 10^{-4.2} - 10^{-3.3}$, averaged over an octave of the dominant
wavenumber and assuming a turbulent region about a parsec in size.
It is not yet clear how to draw a quantitative correspondence between
these inferred observational values and the fluctuation levels seen
in our computations, since the latter correspond to local values and
must be spatially averaged over wavenumber and over the distance from
the supernova remnant shock front. However, the large amplitudes
which we observe in our simulations suggest that the mechanism
considered in this paper may in principle be able to produce averaged
amplitudes comparable to those inferred for the high-turbulence
clumped component of the interstellar medium.

The above discussion is quite preliminary in the context of actual application of our computational results to the astronomical observations of scintillation. Nevertheless, we suggest that Alfven-wave turbulence near supernova remnant shock waves is capable of driving acoustic turbulence to an amplitude appropriate for that component of the interstellar medium that is inferred from radio scintillation measurements to be highly turbulent and clumped. The resulting sound-wave turbulence has short wavelength, and feeds energy into the high-k components of the overall turbulence spectrum.

To make a convincing astronomical application of these ideas, the following issues will have to be addressed: How large is the spatial region containing the strong turbulence, in this model? How do our results scale for other values of the ratio of magnetic to thermal energy density in the interstellar medium? How do our results scale with differing values for the (poorly known) lower cut-off for the cosmic-ray distribution function? And finally, we need to gain a more quantitative understanding of the mechanisms which damp out the sound-wave energy in our code and in the various phases of the interstellar medium, and of how these damping mechanisms depend on the wavelength of the sound waves.

ACKNOWLEDGEMENTS

This research was supported by the US Department of Energy under contract No. W-7405-Eng-48 to the Lawrence Livermore National Laboratory.

REFERENCES

1. J.M Cordes, J.M. Weisberg, and V. Boriakoff, Astrophys. J. **288**, 221 (1985).
2. A.R. Bell, Mon. Not. Roy. Ast. Soc. **182**, 147 and 443 (1978).
3. R.D. Blandford and J.P. Ostriker, Astrophys. J. (Lett.) **221**, L229, (1978).
4. A.L Zachary and B.I Cohen, J. Comput. Phys. **66**, 469 (1986).
5. A.L. Zachary, "Resonant Alfven Wave Instabilities Driven by Streaming Fast Particles", Ph.D. Dissertation, Univ. of California, May, 1987.

SHOCK-ASSOCIATED MHD WAVES: A MODEL FOR INTERSTELLAR DENSITY FLUCTUATIONS

Steven R. Spangler
University of Iowa, Iowa City, Iowa 52242

ABSTRACT

We discuss the possibility that the density fluctuations responsible for radio scintillations could be due to ion-beam-generated magnetohydrodynamic waves near interstellar shock waves. This suggestion is inspired by spacecraft observations which reveal these phenomena near shocks in the solar system. The model quite naturally accounts for the scale on which these fluctuations occur; it is dictated by the wavelength of the unstable waves. A feature of the model is that there is no cascade from very large (of order parsec) scales to small scales. The fluctuations are directly produced on the small scales where they have a pronounced effect on radio waves. An embarrassment to the model is the apparently small spatial extent of regions of intense MHD waves and associated density fluctuations. Simple estimates based on quasilinear theory yield regions too thin to produce the observed scattering. A number of ways in which this conclusion can be circumvented are discussed.

INTRODUCTION AND A SOLAR SYSTEM ANALOG

A major question in the study of interstellar turbulence is the mechanism by which density fluctuations are generated on scales ($10^9 - 10^{13}$ cm) which are very small compared to the global scales in the interstellar medium.

We believe that observational and theoretical studies of phenomena in the solar system can be of considerable assistance in this regard. Research in the last ten years has shown that similar plasma phenomena are seen in the vicinities of many different shock waves. These shocks are

- The Earth's bow shock.

- Propagating interplanetary shocks.

- Planetary bow shocks of Venus, Jupiter, and Saturn.

- Regions of interaction between the solar wind and cometary material.

Observations show that the regions in front of these shocks, termed *foreshocks*, are characterized by reflected ions and large amplitude Alfven waves driven unstable by the ions. Large amplitude density fluctuations accompany these Alfven waves. Spacecraft observations of density fluctuations in the vicinity of the Earth's bow shock are shown in Figure 1.

A recent study[1] has been made of the density fluctuations in the vicinity of the Earth's bow shock, and reached the following conclusions about the density fluctuations produced by these shock-associated Alfven waves.

- The density fluctuations are large, with $\frac{\sigma_n}{n_0} \simeq 0.15$.

Physics of Scattering Media

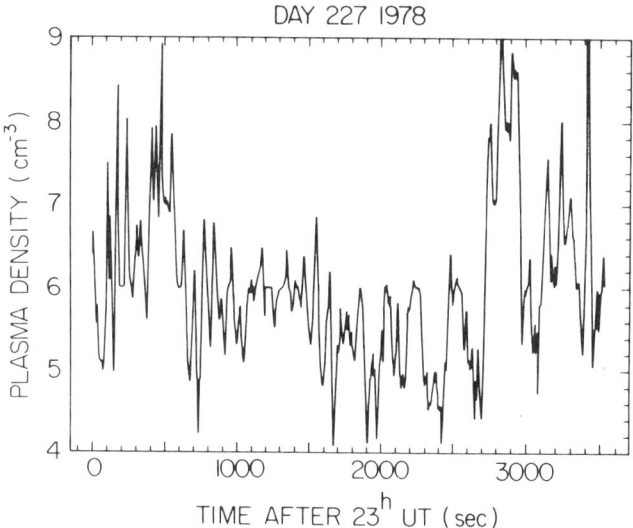

Figure 1: Plasma density fluctuations associated with magnetohydrodynamic waves in the region upstream of the Earth's bow shock. Observations from reference (1).

- The scale for the density fluctuations ranges from the wavelength of the Alfven wave to perhaps 10-15 times that value.

- The primary mechanisms for generation of density fluctuations appear to be oblique propagation, with $\theta_{\vec{k}\cdot\vec{B}}$ of order a few degrees, and nonlinear, ponderomotive modification of the plasma density.

APPLICATION TO THE INTERSTELLAR MEDIUM

The above section discusses observations that Alfven waves, produced by streaming particles associated with shock waves in the solar system, produce large amplitude density fluctuations on scáles much smaller than "global" scales in the interplanetary medium.

It seems virtually certain that the interstellar medium abounds with shock waves, so it is likely that the phenomena described above occur there as well. Do the resultant density fluctuations account for the turbulence affecting radio signals?

Due to the inhomogeneity of the interstellar medium, we must first ask what phase of the medium would be likely to support these fluctuations. The McKee-Ostriker hot phase, or regions such as the local x-ray emitting bubble can be excluded. The reason for this is that with $B \sim 5 \times 10^{-6}$ Gauss, $n_e \sim 4 \times 10^{-3} cm^{-3}$ and $T \sim 10^6$ K, the Alfven speed would be of the order of 200 km/sec, while the acoustic speed would be of the order of 100 km/sec. Such high speeds for the irregularities appear excluded by the observed correlation between pulsar proper motions and scintillation fading times[2,3].

This relationship indicates that pulsar motion dominates the pulsar-turbulence relative motion, and seems to exclude the possibility that the irregularities could move as fast as 100 km/sec.

We thus concentrate on the McKee-Ostriker warm intercloud medium (WIM) or the warm neutral medium (WNM). These media may be characterized by $B \sim 5 \times 10^{-6}$ Gauss, $n_e \sim 0.4 cm^{-3}$, $T \sim 10^4$ K, and a filling factor of perhaps 40 %. The Alfven and ion acoustic speeds are 20 and 10 km/sec, respectively, which are consistent with the pulsar proper motion measurements.

What is the scale on which shock-generated waves would be produced? The waves are generated by a cyclotron resonance with the beam ions, which if $\omega \ll \Omega_i$, gives the following expression for the wavelength of the generated waves

$$\lambda \approx \frac{2\pi v_b}{\Omega_i} \qquad (1)$$

where Ω_i is the ion cyclotron frequency and v_b is the speed of the streaming particles. Observations in the solar wind indicate that $v_b \sim 2 - 3V_s$, where V_s is the speed of the shock wave. If we let $V_s \equiv MV_A$, where V_A is the Alfven speed with the value given above, and use the empirical relation that density fluctuations are produced on scales from 1 to 15 times the wavelength, we have the following estimate for the scales on which the fluctuations are generated;

$$l \sim (10^9 - 10^{10})M \ cm \ . \qquad (2)$$

For Alfvenic Mach numbers of even a few, density fluctuations on the observed diffractive scales would be produced. Such shock speeds could be produced by old supernova remnants, which are doubtlessly common in the interstellar medium, as well as many other possible mechanisms.

The next question to be addressed is whether such foreshocks are sufficiently large to produce detectable scattering. Quasilinear theory indicates that the thickness of the foreshock should be a few times the following length scale

$$l_Q \simeq \frac{D_Z}{V_s} \qquad (3)$$

where D_Z is the spatial diffusion coefficient. With a simple model for D_Z, we have

$$l_Q \simeq \frac{v^2}{V_s \Omega_i}\left(\frac{B}{b}\right)^2 \eta \qquad (4)$$

where v is the speed of the ions, B and b are the static and wave magnetic fields, respectively, and η is the fractional bandwidth of the excited waves. Assuming again that $v \sim 3V_s$, we have

$$l_Q \sim 4 \times 10^8 M \eta \left(\frac{B}{b}\right)^2 \ cm \ . \qquad (5)$$

This size would seem to be far too small to produce the observed scattering. If our suggestion is to be a viable model for interstellar turbulence, then the extent of the scattering region must be much larger than indicated by equation (5). We list below a number of ways in which this might be accomplished.

- If cosmic rays are present in the reflected ion distribution, they will generate waves of longer wavelength, and the foreshock scale will be correspondingly larger.

- The turbulent clumps in which enhanced scattering occurs may contain a very large number of such shocks.

- Processes may be operative which invalidate the above quasilinear expression for the spatial diffusion coefficient.

- Some mechanism other than shock reflection of particles might be responsible for the generation of the MHD waves. For example, Alfven waves near comets are produced by neutral radicals which are photoionized, thereby creating an unstable particle distribution. The extent of such a region is determined by the mean free path for photoionization, not a quasilinear diffusion scale.

ACKNOWLEDGEMENTS

This research was supported at the University of Iowa by grants NAGW-806 and 831 from the National Aeronautics and Space Administration.

REFERENCES

1. S. Spangler, S. Fuselier, A. Fey, and G. Anderson, J. Geophys. Res., 93, 845, (1988).

2. A.G. Lyne and F.G. Smith, Nature, 298, 825 (1982).

3. J.M. Cordes, Astrophys. J., 311, 183, (1986).

HYDROMAGNETIC WAVE HEATING OF LOW DENSITY INTERSTELLAR GAS

Ellen G. Zweibel, Katia M. Ferriere, and J. Michael Shull
University of Colorado at Boulder

I. Introduction

This paper briefly considers the generation and dissipation of compressive waves in the interstellar medium. A more comprehensive version is submitted for publication elsewhere[1], with emphasis on heating the warm, low density phase of the interstellar gas by wave dissipation. In the paper we place our results in the context of radio wave scattering. We have two goals in this regard: to constrain theories of the origin of small scale density irregularities in the interstellar medium, and to make some inferences about the properties of the gas in which the scattering occurs.

We assume that in many respects the interstellar medium is as described by McKee and Ostriker.[2] Hot coronal gas occupies a large volume of galactic disk. Cold clouds are embedded within this hot medium. Warm neutral gas (T $\approx$ 8 x 10^3 °K, $n_H \approx$ 0.2) exists in relatively small puffs more or less surrounded by the hot coronal phase.

Our theory is based on these assumptions as follows. Supernova explosions are taken to be the dominant source of wave energy. Although Type II supernovae may explode predominantly in OB clusters associated with fairly dense gas, we assume that about half of Type I supernovae occur in the hot coronal gas and produce shock waves which expand to very large radii. These shock waves encounter clouds, generating acoustic waves according to the mechanism originally proposed by Spitzer[3,4]. The waves propagate through the hot gas and impinge on warm gas, which for the purposes of numerical estimation is assumed to be bathed in an isotropic flux of waves. Thus, while our work does not require strict adherence to the McKee-Ostriker model, some of the geometric assumptions as well as the phase properties of the gas are assumed to be similar.

Section II of this paper describes our calculations of wave generation. Section III discusses the wave dissipation (or heating) rate in the warm neutral gas. Section IV is a discussion of the radio wave scattering problem.

II. Origin of the Wave Spectrum in the Hot Gas

Our calculations of the wave spectrum has two parts. We first calculate the spectrum of waves generated by encounters of supernova remnants with clouds. These waves subsequently undergo both linear dissipation and nonlinear steepening as they propagate through the hot gas. In principle we might imagine that an energy cascade over many orders of magnitude results from nonlinear processes. We find, however, that viscous damping dominates nonlinear steepening at a fairly low wave number, suggesting that such a cascade does not occur.

A. Generation of Waves

The production of acoustic waves by encounters between an isolated cloud and a supernova shock wave has been considered by Spitzer [3,4] and we only summarize known results. When the shock sweeps over the cloud, a reflected shock will be generated, and if M_s, the Mach number of the incident shock, is less than 2.76, this reflected shock will weaken into an acoustic pulse (if $M_s > 2.76$, a stationary bowshock forms). The wavelength of the pulse generated by a cloud of radius a is about 5a. The energy radiated by the cloud is

$$E_a = \frac{9\pi a^3}{16} \rho_h c_{sh}^2 \zeta \frac{(M_s^2-1)^2}{M_s^2+3} \tag{1}$$

(the subscript h refers to the hot medium). The constant ζ is of order unity. The power radiated by the cloud is therefore

$$P_a = 4\pi \int_{R_m}^{R_M} dR_s \, R_s^2 \, E_a \, r_{sn} \tag{2}$$

where r_{sn} is the supernova rate per unit volume in the hot medium, R_M and R_m are the radii at which $M_s = 1$ and $M_s = 2.76$, respectively, and R_s is the shock radius, to be treated as a function of M_s.

We have evaluated P_a separately for supernovae of Types I and II. If we assume that the Type II supernovae occur in OB associations and are spatially correlated then the resulting supershells make a relatively small contribution to P_a. The numerical value of P_a for Type I supernovae is found to be

$$P_a \approx 6.9 \times 10^{29} \text{ erg s}^{-1} (a_{pc})^3 \, \chi \, f_{50} \tag{3}$$

Here, f_{50} is the Type I supernova rate in units of one per 50 years and χ is a parameter of order unity which depends on the assumed expansion law; in particular $\chi = 1$ for the standard Sedov-Taylor expansion.

The total acoustic wave power P radiated into the interstellar medium is obtained by integrating equations (2) over a spectrum of clouds. If we assume the power law cloud spectrum chosen by McKee and Ostriker, we find that

$$P \approx 1.8 \times 10^{-27} \text{ erg cm}^{-3} \text{ s}^{-1} \, \chi \, f_{50} \tag{4}$$

The input of power is dominated by the largest clouds.

B. Propagation of Waves

The acoustic pulse emitted by a shocked cloud spreads geometrically, steepens nonlinearly (generating power at short wavelengths), and undergoes viscous damping. The first process is independent of wavelength while the second and third occur at rates inversely proportional to the wavelength and the square of the wavelength, respectively, so that viscosity dominates at short wavelengths. Geometrical divergence reduces the wave amplitude and therefore delays steepening.

As noted before, the waves by clouds have wavelengths about 5 times the cloud radius. The clouds have radii between about 2.1pc and 10.8pc, so the corresponding range of wave numbers is 3.8×10^{-20} cm^{-1} to 1.9×10^{-19} cm^{-1}. It turns out that over this range, the steepening rate is less than the viscous damping rate. Therefore we consider the development of a cascade of power to short wavelengths to be unlikely. Rather, the energy density in waves emitted by a single cloud is simply given by

$$W_a(r) = \frac{P_a \exp[-2\gamma_d(r-a)/c_{sh}]}{4\pi r^2 c_{sh}} \quad (5)$$

where γ_d is the viscous damping rate.

III. Heating of the Warm Neutral Gas

Magnetohydrodynamic waves in partially ionized gas are dissipated by ion-neutral friction. The damping heats the gas, so that if $\omega_i(k)$ is the wavelength dependent imaginary part of the wave frequency and $W(k)$ is the energy density in wave of waves number k then the heating rate is

$$H = \int d^3k \, 2\omega_i(k) \, W(k) \quad (6)$$

There are two wave propagation regimes in a partially ionized medium, distinguished by the degree to which ions and neutrals are coupled[5]. In the short wavelength regime, coupling is weak and the phase speed of the waves is essentially determined by the ionized component alone. In the long wavelength regime, coupling is strong and the ionized and neutral species determine the wave speed. We define the collision frequencies ν_{in} and ν_{ni} which measure respectively the rate of momentum transfer from ions to neutrals and neutrals to ions (the relationship $\rho_i \nu_{in} = \rho_n \nu_{ni}$ follows from momentum conservation). We also define the Alfven speeds in the ionized and combined fluids $V_{Ai} \equiv B/\sqrt{4\pi\rho_i}$; $V_A \equiv B/\sqrt{4\pi(\rho_i + \rho_n)}$. In this notation the long wavelength regime is delimited by $k < 2\nu_{ni}/V_A$ and the short wavelength regime by $k > \nu_{in}/2V_{Ai}$. (In the interval between these two limits, the waves do not propagate).

When the waves generated by cloud-supernova shock encounters propagate from the hot to the warm gas they suffer partial reflection and change in wavenumber. If the interface is assumed to be sharp, the transmission coefficient T_r is of order $(\rho_h/\rho_w)^{1/2} \approx .12$ while the wave numbers in the hot and warm gases k_h and k_w are related by $k_w \approx k_h (\rho_w/\rho_h)^{1/2}$. Taking this into account we find that these waves are in the long wavelength regime and are damped at a rate

$$\omega_i = k^2 v_A^2 / 2\nu_{ni} \quad (7)$$

(We[1] have also included the effects of thermal pressure, which results in a small correction to equation (7)).

We evaluate the heating rate by first integrating equation (5) over all cloud positions, weighing it by the cloud size distribution function, and correcting for partial reflection. This yields $W(k)$. We then perform the integral in equation (6), using equation (7). The result is

$$H \approx 4.2 \; 10^{-27} f_{50} \chi \; \text{erg cm}^{-3} \text{s}^{-1} \tag{8}$$

Which is to be compared with a cooling rate[6] $L \approx 7.7 \; 10^{-28}$ erg cm^{-3} s^{-1}. If we set $f_{50} \approx 0.5$ and take for χ the value 0.47 corresponding to McKee and Ostriker's model of supernova remnant evolution then H exceeds L by 28%. Given the many uncertainties in the problem we consider it significant and H and L are fairly close to each other, and also that we err on the side of predicting too much heating.

Despite the arguments of Section II B, we have also considered the consequences of assuming that the energy density in waves generated by supernovae is distributed in a Kolmogorov spectrum down to scales of 10^{11} cm[7]. The short wavelength waves are damped at the rate

$$\omega_i = \nu_{in}/2 > k^2 v_A^2/2 \nu_{ni} \tag{9}$$

and the resulting heating rate exceeds the cooling rate L by about three orders of magnitude. Similar conclusions have been reached before[8].

IV. Implications for Radio Wave Scattering

Two pertinent points about the radio wave scattering problem emerge from our work. The first concerns the source of small scale density fluctuations in the interstellar medium. Substantial power in waves is generated by encounters between supernova shock waves and clouds embedded in the hot coronal gas. But we have shown that these waves, which are emitted at wavelengths ordered by the cloud size, suffer viscous decay before they steepen appreciably. Thus little power reaches short wavelengths. The second point concerns the thermal balance of the medium in which scintillation occurs. If this medium is the warm, weakly ionized gas and the density fluctuations resemble hydromagnetic waves in a Kolmogorov spectral energy distribution then the wave dissipation rate corresponds to a heating rate some 10^3 times the cooling rate. It may be simplest to conclude that the warm neutral gas is not a site where scintillation occurs.

Acknowledgements: Support for this research was provided by NASA Grants NAGW-91 and NAGW-766 to the University of Colorado.

References

1. K.M. Ferriere, E.G. Zweibel, and J.M. Shull, submitted to Ap. J. (1988)
2. C.F. McKee and J.P. Ostriker, Ap. J. 218, 148 (1977)
3. L. Spitzer, Ap. J. 262, 315 (1982)
4. S. Ikeuchi and L. Spitzer, Ap. J. 283, 825 (1984)
5. R.M. Kulsrud and W.A. Pearce, Ap. J. 156, 445 (1969)
6. J.M. Shull and D.T. Woods, Ap. J. 288, 50 (1985)
7. L.C. Lee and J.R. Jokipil, Ap. J. 204, 735 (1976)
8. C.J. Cesarsky, Ann. Rev. Astron. Astrophys. 18, 289 (1980)

Oceanic and Interstellar Fossil Turbulence

Carl H. Gibson
University of California at San Diego, La Jolla, CA 92093

Abstract

Turbulence in the stratified ocean is inhibited at large scales by gravity, leaving internal waves and partially mixed fluctuations in the density field called fossil turbulence. Information about previous turbulence activity is preserved by the fossil remnants of 3-D and possibly 2-D turbulence. Turbulence in the interstellar medium may be inhibited by gravity at large scales, and by self gravity and electromagnetic forces at small scales, also producing fossils of previous active turbulence.

Introduction

Motions of the interstellar medium are extremely complex and poorly understood. Velocity differences of order 10^7 m/s exist on galactic scales up to 10^{21} m. Atomic densities range from 10^{12} to 10^3 m^{-3} with nearly constant energy density of 10^{-19} kg·m^{-1}·s^{-2}. Magnetic field strengths are of order 10^{-10} T, but may be concentrated by gravitational collapse. Much of the matter is ionized gas in a state of hypersonic, magnetohydrodynamic turbulence, shocked by supernovas every few decades. Preferred temperatures are 10^2, 10^4 or 10^6 K. Under the circumstances one might expect the results of incompressible turbulence and turbulent mixing studies to have limited or no applicability to the description of the dynamics of the interstellar medium. On the other hand one is faced with the intriguing fact, shown by Fig. 1 of Armstrong et al.[3], that the spectrum of density fluctuations in the interstellar medium obeys the Oboukhov-Corrsin[1,2] $q^{-5/3}$ wavenumber spectrum over a range of wavelengths $\lambda = 2\pi/q$ from about 10^{10} to 10^{19} m, suggesting some sort of turbulent mixing process may be, or have been, at work. This nine decade range of length scales from stellar to galactic disk thickness matches the viscous to planetary millimeter to megameter range of turbulence scales which occur in the ocean. The purpose of the present note is examine possible similarities in the dynamics of turbulent processes on oceanic and cosmic scales, particularly the tendency of turbulence damped by Coriolis or gravity forces to leave behind remnant fluctuations in various fields, termed fossil turbulence. It is speculated that the size of the galaxy may reflect a fossil Rossby scale of 2-D turbulence and the size of the solar system may reflect a fossil Ozmidov scale of 3-D turbulence.

Physical Processes

Turbulent motions arise when the nonlinear inertial forces of the momentum conservation equation are not damped by other resulting forces. The criterion for the existence of turbulence is that the ratios of inertial to viscous, inertial to buoyancy, inertial to Coriolis, or inertial to magnetic forces must be greater than critical values. These ratios are the Reynolds number $Re = UL/\nu$, the inverse Richardson number $Ri^{-1} = U^2/N^2$, the Rossby number $Ro = U/fL$ and the inverse square of the Alfen number $A^{-2} = \rho\mu'U^2/B^2$, respectively, where U is a characteristic velocity difference on length scale L, ν is the kinematic viscosity, N is the intrinsic frequency of a stably stratified fluid $\equiv (g\rho^{-1}\partial\rho/\partial z)^{1/2}$, g is the acceleration of gravity, ρ is the density, f is the vertical component of global rotation, μ' is the permeability (kg·m/A^2·s^2), and B is the magnetic field strength (tesla = kg/A·s^2). Inertial forces $F_I = \rho U^2 L^2$, viscous forces $F_V = \rho \nu U L$, Coriolis forces $F_C = \rho f U L^3$ and magnetic forces are $F_L = B^2 L^2/\mu'$. In a stably stratified fluid like the ocean with constant vertical density gradient the buoyancy forces $F_B = g\Delta\rho L^3 = \rho N^2 L^4$. In the interstellar medium gravity is not constant but depends on local concentrations of mass which can increase if not mixed by turbulence. When turbulence levels fail to prevent mass accumulation, buoyancy forces can develop which locally fossilize turbulence and accelerate the condensation process.

Turbulence is the result of shear instabilities. The rate of change of momentum per unit volume is $\rho[\partial v/\partial t + (v\cdot\nabla)v] = \rho[\partial v/\partial t + \nabla(v^2/2) - v\times\omega]$, where ω is the fluid vorticity $\omega \equiv$ curl v. Vortex sheets are unstable to all perturbations because vortex forces $v\times\omega$ arise in the direction of the perturbation to cause amplification and the formation of eddy-like motions. The eddies grow and acquire energy by entrainment. Generally the smallest instabilities have the smallest time scale, so turbulence begins at the smallest scale permitted by viscous, buoyancy, Coriolis or magnetic forces and "cascades" to larger scales by multiple vortex pairings. Note that this model of the turbulence cascade is contrary to conventional wisdom derived from the Richardson poem ("big whorls have little whorls that feed on their velocity, and smaller whorls have smaller whorls, and so on to viscosity") that the big eddies form first and the small eddies form later. Consequently there is no difference between the directions of the turbulence cascades for 2-D versus 3-D turbulence, as often supposed. Vortex sheets (shear layers) are one-dimensional, eddies are two-dimensional and become three-dimensional unless constrained by gravity or electromagnetic forces.

In the following, active turbulence is defined as the eddy-like state of fluid motion that arises when the inertial forces of the eddies are larger than viscous, buoyant, Coriolis, magnetic or any other forces which tend to damp out the eddies. Fossil turbulence is defined to be a fluctuation of any hydrophysical field such as the density, temperature, concentration or vorticity produced by active turbulence which persists after the fluid is no longer actively turbulent on the scale of the fluctuation.

Most of the ocean is stably stratified, meaning that density increases in the direction of gravity. Turbulent motions are therefore opposed by gravity since turbulent eddies try to force light fluid down and bring heavy fluid up. Shear layers are produced by currents and internal waves on sheets of maximum N and are stabilized by buoyancy. Turbulence occurs rarely in growing isolated patches with maximum ε and minimum size. The maximum size is limited by buoyancy forces. From the Kolmogoroff-Oboukhov law $U = U(\varepsilon, L) \approx (\varepsilon L)^{1/3}$ so the inertial forces of the turbulence are balanced by buoyancy forces at the Ozmidov scale $L_R \equiv (\varepsilon/N^3)^{1/2}$ and fossilization begins. The energy scale L_O of the turbulence first increases till $L_O \approx L_R$, then decreases with L_R until viscous forces become important at the inertial-viscous or Kolmogoroff length scale $L_K \equiv (\nu^3/\varepsilon)^{1/4}$. L_O then ceases to exist as the turbulence vanishes. This buoyant-inertial-viscous point is called the point of complete fossilization and occurs when $L_O \approx L_R \approx L_K \approx L_{KF} \equiv (\nu/N)^{1/2}$. The buoyant-inertial-viscous scale L_{KF} is termed the fossil Kolmogoroff scale[4].

During the fossilization process entrainment continues, but at a decreasing rate. Most of the kinetic energy of the turbulence at the point of fossilization is preserved as saturated internal wave motion at length scale L_{Ro}, where the o-subscript indicates the inertial-buoyancy point, or "point of fossilization". Various overturn scales of the microstructure can be defined, such as the microstructure patch height, which do not decrease as rapidly as L_R but remain constant at about L_{Ro}, thus preserving information about previous turbulence activity, particularly the previous dissipation rate ε_o. This ability to preserve information is the most useful aspect of fossil turbulence, just as it is for other fossils.

Fossil turbulence was first noticed from persistent refractive index fluctuations detected in the wake of mountains by radar scattering, and from temperature microstructure patches devoid of detectable velocity fluctuations by a submarine in the ocean[4]. Spectral modelling and estimates of universal constants[5] have been compared to oceanic[6,7,8] and laboratory tests[9,10] with satisfactory agreement, although it should be mentioned that the inference that most ocean microstructure is fossil turbulence is not presently accepted by all oceanographers[11,12,13]. It is easy to observe fossil turbulence formation by pouring cold milk in warm coffee: the initial turbulent patch grows, is damped by gravity, and leaves a bobbing fossil vorticity turbulence wave motion of the partially mixed fossil milk turbulence. Sky writing and jet contrails become fossil turbulence when they occur above the inversion layer.

Another class of fossil turbulence occurs for 2-D turbulence. Satellite images of the ocean surface temperature show a wide range of eddy scales exist, or have existed, at scales up to hundreds of kilometers in diameter. The mixing patterns are identical to those obtained by slicing a plane from a 3-D turbulent mixing system, showing the familiar vortex pair structures over a range of scales. The constraint at large scales is the Coriolis force $2\Omega \times v$, where Ω is the earth's rotation rate. Equating inertial and Coriolis forces for 2-D turbulence with dissipation rate ε gives a maximum Rossby scale $L_\Omega \equiv (\varepsilon/f^3)^{1/2}$, where f is the vertical component of Ω. Thus eddies near the equator tend to be larger than those at high latitudes. Because ε will generally decrease, the 2-D turbulence scale will decrease leaving saturated gyroscopic-intertial-gravity wave motions and partially mixed, remnant 2-D turbulence fluctuations in stirred scalar fields such as temperature. Just as 3-D turbulence fossils preserve information about previous active 3-D turbulence, 2-D turbulence fossils such as $L_{\Omega o}$ may preserve information about previous active 2-D turbulence. Little is known of 2-D turbulence at present, let alone 2-D fossil turbulence.

Cosmic Scale Turbulence

Consider the dissipation rate of turbulence in the universe ε_U. Assuming the initial energy per unit mass is some fraction of c^2 from Einstein's equation, where c is the velocity of light 3×10^8 m/s, and the age of the universe $\tau_U \approx 10^{17}$ s we find

$$\varepsilon_U \approx c^2/\tau_U \approx (3 \times 10^8)^2/10^{17} \approx 1 \text{ m}^2/\text{s}^3$$

at the present time. This may be compared to the dissipation rate in the galaxy estimated from the Kolmogoroff-Oboukhov law using a characteristic velocity difference $v \approx 10^7$ m/s and a galactic length scale $L_G \approx 10^{21}$ m

$$\varepsilon_G \approx v^3/L_G \approx (10^7)^3/10^{21} \approx 1 \text{ m}^2/\text{s}^3$$

which seems a rather remarkable coincidence, both in the equality of the two dissipation rates and the order one value of $\varepsilon_U \approx \varepsilon_G$ in SI units. By this model, the galaxy would appear to be actively turbulent at the largest scales.

If we take the galactic disk size L_G to be a 2-D fossil Rossby scale L_{Ω_0} of a rotating, expanding, gravitationally dominated system, this implies a rotation rate at fossilization $\Omega_0 \approx 10^{-14}$ rad/s, or a period of twenty million years. If the galactic material separated from the larger system after fossilization, the frictional forces causing it to rotate with the system would cease, along with the gravitational forces constraining the motions to 2-D.

Fossil turbulence in the galactic "fluid" can be expected to occur on small scales $L_{Ro} \ll L_O \approx L_G$ rather than at the turbulent energy scale $L_{Ro} \approx L_O$ as occurs in the ocean. The reason is that the gravitation in the ocean is uniform so buoyancy forces increase with vertical distance in the stratified fluid. Gravity in the interstellar medium, however, is a local process self induced by concentrations of mass. Supernova blast waves concentrate mass but also induce strong turbulence which will tend to keep the density homogeneous and resist gravitational collapse. Eventually radiative cooling and internal friction will reduce the turbulence kinetic energy u^2 and ε in such regions until gravity from density concentration or particles can overcome viscous forces and begin to collect mass. If a particle (or density concentration) of mass M is imbedded in a viscous flow of fluid with density ρ and "viscosity" μ (viscosity in such fluids may involve electromagnetic forces) with rate-of-strain $\gamma \equiv (\varepsilon/\nu)^{1/2}$, the viscous forces $F_V = \mu \gamma L^2$ will balance gravitational forces $F_g = \rho GML$ at length scale L_g

$$L_g \equiv GM/\nu\gamma$$

where $G = 6.67 \times 10^{-11}$ N·m^2/s^2 is the universal gravitational constant. When L_g becomes larger than the size of the particle, gravitational collapse can begin. As the nucleus grows, M increases, further increasing L_g until it equals the Kolmogoroff scale L_K. For $L_g > L_K$ the extent of the region affected by gravitational forces changes to

$$L_{g'} \equiv (GM)^{3/5}/(\varepsilon)^{2/5}$$

and the mass collection continues. When the rate of mass collection becomes so large that the assumption of constant density of the medium fails then stable stratification develops as N increases, with the denser fluid near the nucleus. The Ozmidov scale $L_R \equiv (\varepsilon/N^3)^{1/2}$ decreases as $L_{g'}$ increases. When they are equal, fossilization of the turbulence begins, with $\varepsilon_0 = N_0^{5/3}(GM)^{2/3}$ and

$$L_{g'0} = L_{R0} \equiv (\varepsilon_0/N_0^3)^{1/2} = (GM/N_0^2)^{1/3}$$

where N_0 is the intrinsic frequency at fossilization. Dissipation rates near the nucleus decrease as the turbulence is damped by buoyancy forces and friction and the entrainment rate of material from the surroundings decreases. Angular momentum and magnetic field lines collected by gravitational collapse on scale L_{R0} will be preserved as fossil remnants of the ambient values at the time of beginning fossilization. See Scalo[14] for a review of other theoretocal approaches to interstellar turbulence.

Summary

The preceding turbulence modeling is quite preliminary and speculative, but demonstrates the possibility that fossilization effects may be relevant to dynamics on galactic or even larger scales just as they are in the ocean and atmosphere.

Acknowledgements

The author wishes to acknowlege several useful conversations on the material presented here with participants at the conference, particularly with Barney Rickett. Funding was provided by ONR Contract N00014-85-C-0104.

References

1. Oboukhov, A. M., Structure of the temperature field in turbulent flows. Izvestiya Akademii Nauk SSSR, Georgr. and Geophys. Ser. **13**, 58 (1949).
2. Corrsin, S., On the spectrum of isotropic temperature fluctuations in isotropic turbulence. Journal of applied Physics **22**, 469 (1951).
3. Armstrong, J. W., J. M. Cordes and B. J. Rickett, Density power spectrum in the local interstellar medium. Nature **291**, 561-564 (1981).
4. Woods, J. D. Ed., Report of working group (V. Hogstrom, P. Misme, H. Ottersten, and O. M. Phillips): fossil turbulence. Radio Science **4**, 1365-136 (1969).
5. Gibson, C. H., Fossil temperature, salinity and vorticity turbulence in the ocean. in *Marine Turbulence*, J. C. J. Nihoul Ed., Elsevier, N. Y., 221-257 (1980).
6. Gibson, C. H., Internal Waves, Fossil Turbulence, and Composite Ocean Microstructure Spectra. J. Fluid Mech. **168**, 89-117 (1986).
7. Gibson, C. H., Fossil turbulence and intermittency in sampling oceanic mixing processes. J. Geophys. Res. **92**(C5), 5383-5404 (1987).
8. Gibson, C. H., Oceanic turbulence: big bangs and continuous creation. J. Physicochem. Hydrodyn. **8**(1), 1-22 (1987).
9. Stillinger, C. C., M. J. Head, K. N. Helland, and C. W. Van Atta, Experiments on the transition of homogeneous internal waves in a stratified fluid. J. Fluid Mech. **131**, 91-122 (1983).
10. Itsweire, E. C., K. N. Helland and C. W. Van Atta, The evolution of a grid-generated turbulence in a stably stratified fluid. J. Fluid Mech. **162**, 299-338 (1986).

11. Dillon, T. R., The energetics of overturning structures: implications for the theory of fossil-turbulence. J. Phys. Oceanogr. **14**, 541 (1984).
12. Caldwell, D. R., Oceanic turbulence: Big bangs or continuous creation? J. Geophys. Res. **88**(C12), 7543-7550 (1983).
13. Gregg, M. C., Diapycnal mixing in the thermocline: A review. J. Geophys. Res. **92**(C5), 5249-5286 (1987).
14. Scalo, J. M. Theoretical approaches to interstellar turbulence. in *Interstellar Processes (Astrophysics and space science library 134)*, D. J. Hollenback and H. A. Thronson, Jr. Eds., D. Reidel Pub. Co., Dordrecht, Holland, 349-392 (1987).

III. TURBULENCE IN THE SOLAR WIND

Radio Observations of Plasma Irregularities in the Solar Wind

A. Hewish

Cavendish Laboratory, Madingley Road,
Cambridge, CB3 0HE, U.K.

ABSTRACT

Ground-based observations of radio wave scattering caused by density irregularities in the solar wind provide a useful means of studying the inner heliosphere. This review outlines the different methods that have been used and highlights some of the important results obtained.

1. REFRACTIVE SCATTERING NEAR THE SUN

Historically, the first attempts to detect electrons by propagation effects in the outer corona were made in the early 1950s following the discovery of radio sources, in particular the Crab Nebula which comes within a few degrees of the sun each year in June. Spherical refraction in the diverging lens formed by the entire ionized corona was expected to produce caustics bounding an occulting disk considerably larger than the sun. In the event, random refraction by small-scale irregularities proved to be larger than systematic angular deviations due to average radial density gradients so that the source was broadened, not occulted. Interferometric observations at metre wavelengths on baselines out to ~10 km enabled angular broadening to be studied out to 0.5 AU from the sun[1]. The scattering was found to be anistropic, with source broadening greatest circumferential to the sun, as would be expected for irregularities tending to form radially-directed filaments. Reductions of source intensity were observed close to the sun; these must occur when the scattered image of a source would exceed the angular extent of a localised scattering region[2].

Both the dominance of small-scale refraction, and the reduction of intensity by a random scattering mechanism under appropriate conditions, may be equally important in the interstellar medium. It should not be assumed that caustics are necessarily involved to explain observed intensity variations.

2. INTERPLANETARY SCINTILLATION

In 1964 it was discovered that radio sources containing hotspots of angular size less than one arcsec illuminated the small-scale irregularities with enough coherence to produce radio speckle patterns at the earth[3]. This causes intensity, scintillation on a typical time-scale of 0.1 - 1.0 Hz and at metre wavelengths the effect is detectable at all solar elongations. This opened the way to investigating the physical scale of the irregularities, the speed of the solar wind by timing the drift of the speckle pattern across the ground, and ultimately to mapping

© 1988 American Institute of Physics

the global structure and movement of large-scale heliospheric disturbances such as corotating interaction regions and interplanetary shocks.

Interplanetary scintillation is remarkably stable when day to day variations are averaged out. If the rms intensity fluctuation at radio frequency ν is Δs, the relation $\nu\Delta s \propto r^{-1.5}$ holds over distances from a few solar radii to more than 1 AU. This radial variation reflects the overall gradient of mean density $N \propto r^{-2}$ and shows that $\Delta N/N$ is nearly constant over the same range. Some mechanism for generating turbulence on scales of 100 - 200 km must exist even when the solar wind has reached its coasting phase.

Much work has been done to elucidate the wavenumber spectrum of the small-scale irregularities. Whether it conforms to a simple Kolmogorov power lay or whether there is some charactersitic scale has been the subject of considerable debate. The problem is that intensity scintillation can only occur at distances exceeding the Fresnel distance, where the scale size of the irregularities is smaller than the radius of the first Fresnel zone. Thus, when a range of scales exists, scintillation only occurs for scales $<(cz/\nu)^{\frac{1}{2}}$, when z is the distance to the irregularities. It is important not to confuse this frequency dependent selection effect with a genuine scale within the scattering medium. The final conclusion is that beyond about 0.3 AU a Kolmogorov spectrum is a reasonable approximation for scales exceeding about 300 km, but the spectrum flattens at smaller scales prior to a cut-off imposed by the proton gyro-radius. Near the sun there are additional complications as the spectrum becomes flatter than Kolmogorov[4,5].

Multi-frequency measurements have been made to investigate the spectrum at scales exceeding the Fresnel scale. Angular refraction caused by large scale irregularities will displace the speckle patterns due to the smaller scales. This effect is dispersive and shows up as characteristic tilts in dynamic spectra where scintillation is displayed in the two-dimensional ν-t plane. Limited data on one radio source at about 50 solar radii showed dynamic spectra remarkably similar to results obtained for pulsars viewed through the interstellar medium[6]. Dual-frequency measurements reveal the same phenomena as time-shifts of intensity variations and more extensive observations have been made. For both types of observation the magnitude of the refraction effects supports a Kolmogorov spectrum for scales an order of magnitude larger than the Fresnel scale[7].

Numerical simulations of diffraction from physically thin phase-screens are a reasonable approximation to reality for lines of sight within elongations of 20° and computations have confirmed that the observed dynamic spectra of intensity scintillation are reproduced by a Kolmogorov model[8].

Scintillation is also observed on coherent transmissions from spacecraft. Dual-frequency phase-locked signals have been especially useful in measurements of phase scintillation. Here the phase variations are, of course, not subject to the Fresnel distance constraint and have confirmed the Kolmorov law at scales exceeding 10^4 km far from the sun. Nearer to the sun the power law slope becomes flatter[9].

3. THE SOLAR WIND

An important advantage of the radio scintillation technique, as compared to in-situ observations using spacedraft, is that it can be used to study conditions at all heliolatitudes and is not confined to regions near the ecliptic plane. It has also given useful information very close to the sun where the solar wind is being accelerated.

Systematic measurements of cross-correlograms of intensity fluctuations at three sites have been carried out by groups at San Diego and Nagoya. The most important finding concerns the increase of solar wind speed at high heliolatitidues. This is related to the polar coronal holes which are most prominent near sunspot minimum. At these times the polar wind flows at roughly twice the speed in the ecliptic plane. Near solar maximum the polar holes shrink and the latitude variation is less dominant[10].

Cross-correlograms have also provided important information on the extent of the acceleration region of the solar wind. Observations at frequencies up to 2.3 GHz have shown that acceleration occurs over a range out to about 30 solar radii. Within this distance there is an increasing random velocity component which dominates at distances less than 5 solar radii. Another method used to observe conditions close to the sun is based on single-site observations of the temporal spectrum of scintillation. For a physically thin scattering screen there must be a sharp cut-off at scintillation frequencies above the Fresnel cut-off. For a power law irregularity spectrum this appears as a characersitic "knee" in the temporal spectrum. In reality the "knee" is blurred because of the finite extension of the scattering zone.

A technique for surmounting this difficulty is the cross-spectrum method. If $M_{11}(f)$ and $M_{22}(f)$ are the temporal spectra observed at frequencies ν_1 and ν_2, and $M_{12}(f)$ the spectrum of the cross-correlation function of the dual frequency intensity variations, it can be shown that the co-spectrum $Re(M_{12})$ $[M_{11} M_{22}]^{-\frac{1}{2}}$ is much more sensitive to the Fresnel cut-off. Application of the method needs a model for the irregularity spectrum but the normalised co-spectrum shows a distinctive cut-off near $f_c = V(zc/\nu_1)^{-\frac{1}{2}}$. A power law spectrum with two slopes gives a reasonable fit to observed co-spectra and the derived solar wind V agrees well with previous estimates within the acceleration region[11],

4. IPS-IMAGES OF INTERPLANETARY TRANSIENTS

When scintillation is observed all over the sky by sampling a grid of 1000-2,500 radio sources on a day to day basis, major transients of plasma density can be mapped and tracked at distances of 0.5 to 1.5 AU[12]. Due allowance must be made for systematic effects, such as the radial gradient of scintillation and the differing coherence of illumination due to the variation of source size. After calibration for these parameters it is possible to detect perturbations of mean plasma density which exceed 20%. Extensive calibration against spacecraft observations has shown that the scintillation enhancement factor, defined as g = Δs (observed)/Δs (expected) for each source, follows the relation g∝$N^{½}$. Contour maps of g (g-maps) thus reveal the 2-dimensional projections of large scale transients against the sky.

Systematic monitoring of transients for over two years during 1978-79 and 1980-81 has produced a wealth of information. During this period near solar maximum the most common transients were bubble-like shells of enhanced density, often associated with shock fronts, which were compression zones driven by fairly sudden eruptions of fast solar wind. In addition, there were corotating interaction zones, due to more stable high-speed streams from coronal holes, which have been thoroughly studied by scintillation methods in the past.

Proton events and geomagnetic activity are closely associated with interplanetary shocks and the use of scintillation to forecast their arrival is currently a topic of considerable interest to agencies such as the Space Environment Laboratory (Boulder) and other prediction services around the world. Steps are being taken to establish more facilities for monitoring many faint sources similar to the 3.6 hectare array at Cambridge. Detailed analysis of g-maps for 1978-79 has given results which contradict the generally accepted solar flare theory of interplanetary shocks. Tracking major shock disturbances back to the sun, using positional data observed in the range 0.5-1.5 AU, has shown that the sources of the eruptions are, in reality, coronal holes at equatorial latitudes[13]. This new insight, due to the removal of much guesswork in assigning solar causes to disturbances at 1 AU, should clarify the complex field of solar-terrestrial physics. However, the solar flare origin of powerful interplanetary shocks is so deeply imbedded that the new conclusions based on g-maps are being hotly contested.

The scintillation of spacecraft transmissions is also being used to study transients by means of "Doppler noise"[14]. This is the spectral broadening of coherent radiation due to scintillation. An advantage of this method is that strong scattering does not lead to saturation when Δs~s. The method can also be used near the sun at frequencies ν~1GHz, but since it relies on spacecraft suitably positioned behind the sun systematic monitoring of transients cannot be maintained over long periods.

REFERENCES

1. A. Hewish, Solar Wind (NASA Washington, D.C. SP-308, 1972) p. 477.
2. A. Hewish, Mon. Not. R. astr. Soc., 118, 238, (1955).
3. A. Hewish, P.F. Scott, and D. Wills, Nature, 203. 1214, (1964).

4. A.C.S. Readhead, M.C. Kemp and A. Hewish, Mon. Not. R. astr. Soc., 185, 207, (1978).
5. J.W. Armstrong, and R. Woo, Astron. Astrophys. 103, 415, (1981).
6. T. Cole and O.B. Slee, Nature, 285, 93, (1980).
7. G.R. Gapper and A. Hewish, Mon. Not. R. astr. Soc., 197, 209, (1981).
8. W.A. Coles and J.P. Filice, Nature, 312, 251, (1984).
9. R. Woo and J.W. Armstrong, J. Geophys. Res, 84, 7288, (1979).
10. M. Kojima and T. Kakinuma, J. Geophys. Res., 92, 7269, (1987).
11. S.L. Scott, B.J. Rickett and J.W. Armstrong, Astron. Astrophys. 123, 191, (1983).
12. G.R. Gapper, A. Hewish, A. Purvis and P.J. Duffett-Smith, Nature, 296, 633, (1982).
13. A. Hewish, and S. Bravo, Solar Phys. 106, 185, (1985).
14. R. Woo and J.W. Armstrong, N.R. Sheeley, R.A. Howard, M.J. Koomen, D.J. Michels and R. Schwenn, J. Geophys. Res. 90, 154, 1985.

THE SOLAR WIND TURBULENCE SPECTRUM NEAR THE SUN

Wm. A. Coles
University of California, San Diego, CA 92093

J. K. Harmon
Arecibo Observatory, Arecibo, PR 00613

ABSTRACT

The results of recent radio propagation observations of the near-Sun solar wind are reported. The new results, derived from radar spectral broadening and VLBI phase scintillations, are compared with earlier observations (angular broadening, spacecraft spectral broadening, intensity scintillation, and phase scintillation) to determine the shape and radial evolution of the power spectrum of plasma density fluctuations. The various data are best reconciled by a turbulence spectrum which is relatively steep (Kolmogorov) at large scales (>10^3 km), has a local flattening at intermediate (10-100 km) scales, and steepens again at an inner (cutoff) scale.

INTRODUCTION

One of the primary goals of radio propagation studies of the solar wind has been to determine the form of the plasma turbulence spectrum over a wide range of spatial wavenumbers. This has been a difficult and challenging task because of the data interpretation problems and limited dynamic ranges inherent in the various probing techniques. The emphasis in recent years has been on the near-Sun region, both because it is an intrinsically interesting region dynamically and because it scatters radio waves sufficiently strongly that it can be probed at centimeter-decimeter wavelengths. The radio techniques which have been used to probe this region include angular and spectral broadening, intensity scintillation, and phase scintillation. Angular and spectral broadening provide estimates of the turbulence spectrum on spatial scales in the range 100 m to 100 km. Spectral broadening measurements are normally made using a spacecraft beacon[1], although we have recently developed a technique for estimating spectral broadening from planetary radar observations[2,3]. Spectrum estimates at large scales (10^3-10^6 km) can be made using the phase scintillation technique. While the most suitable probe for phase scintillation observations is a coherent spacecraft beacon[1], we have recently shown that useful results can also be extracted from VLBI measurements of short-time-scale fluctuations in interferometer visibility phase[3].

From a comparison of the various observational results, Coles and Harmon[3] concluded that the density spectrum in the near-Sun region must depart significantly from a simple power law. These results are summarized here; for a more thorough discussion the reader is referred to Coles and Harmon[3]. Because the data comparison is most easily done in terms of the wave structure function, we will

© 1988 American Institute of Physics

briefly review the theory of the wave structure function before proceeding with a discussion of the interpretation of the observational results.

THE WAVE STRUCTURE FUNCTION

The angular and spectral broadening techniques measure the loss in spatial or temporal coherence suffered by a coherent wave traversing a scattering medium. Interferometer observations of angular broadening involve measurements of the spatial correlation of electric field on a baseline $\vec{s}$ as given by the mutual coherence function $\Gamma(\vec{s}) = <E(\vec{r})E^*(\vec{r}+\vec{s})>/<|E|^2>$. Spectral broadening measures the temporal coherence $\Gamma(\tau)$, which is the Fourier transform of the measured frequency spectrum $P(f)$. This can be converted into a spatial coherence estimate $\Gamma(\vec{s})$ by using a velocity transformation $\vec{s}=\vec{V}\tau$, where V is the drift (solar wind) velocity.

The mutual coherence is directly related to the wave structure function $D(\vec{s})$ through

$$\Gamma(\vec{s}) = \exp\{-D(\vec{s})/2\} \qquad (1)$$

where

$$D(\vec{s}) = <\{\phi(\vec{r})-\phi(\vec{r}+\vec{s})\}^2> . \qquad (2)$$

Here $\phi(\vec{r})$ is the geometric phase delay on a straight-line path from the source to the observer at position $\vec{r}$. For the case where the electron density fluctuation spectrum $\Phi_{ne}(\vec{k})$ is an isotropic function of the transverse wavenumber κ, a wave traversing a slab of thickness Δz will have a wave structure function given by

$$D(s) = 8\pi^2 r_e^2 \lambda^2 \Delta z \int_0^\infty \{1-J_o(\kappa s)\}\Phi_{ne}(\kappa;\kappa_z=0)\kappa d\kappa \qquad (3)$$

where λ is the r.f. wavelength and r_e is the classical electron radius. From equation (3) it follows that a power law spectrum of the form $\Phi_{ne}(\kappa) \propto \kappa^{-\alpha}$ corresponds to a structure function of the form

$$D(s) \propto s^{\alpha-2} \qquad (2<\alpha<4) . \qquad (4)$$

In addition, numerical integration of equation (3) yields two useful approximations; 1) a power law spectrum with a high-wavenumber cutoff at κ_i corresponds to a structure function which breaks from $D(s) \propto s^{\alpha-2}$ to $D(s) \propto s^2$ at an "inner scale" $s_i \cong 3/\kappa_i$, and 2) a two-component power law spectrum with a high-wavenumber inflection or flattening at κ_b corresponds to a structure function with an inflection at $s_b \cong 1/\kappa_b$. Finally, from equation (3) we have the convenient result that structure functions obtained a different wavelengths λ can be compared by simply scaling by the square of the ratio of the wavelengths.

OBSERVATIONAL RESULTS AND DISCUSSION

The structure functions estimated from various observations are shown in Figure 1.

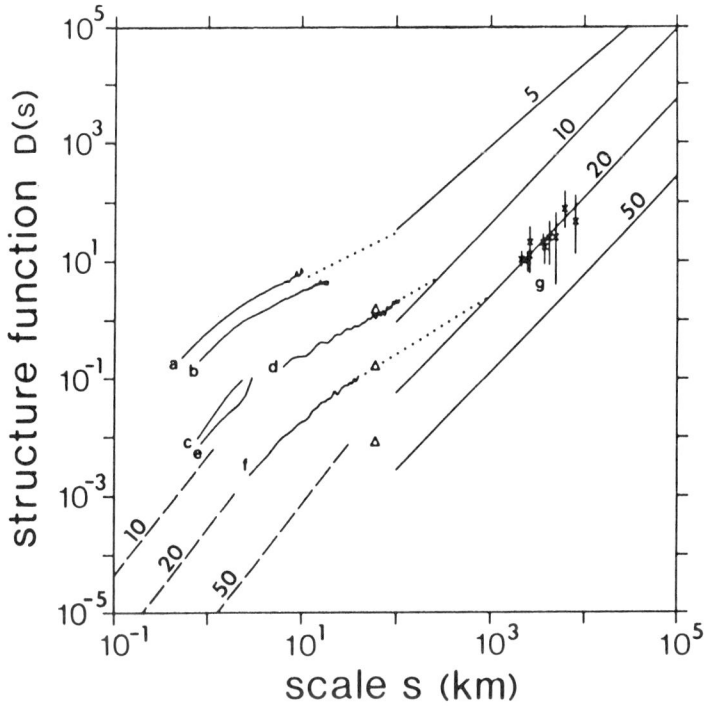

Fig. 1. Structure function estimates obtained near 5, 10, 20 and 50 solar radii from various techniques. See text.

The curves labelled (a)-(f) are a representative subset of functions derived from Arecibo radar spectral broadening observations made in 1981 and 1983 at wavelengths of 2380 MHz and 430 MHz (the latter scaled to 2380 MHz). Curves (a) and (b) were obtained near 5 R_0 (solar radii), curves (c)-(e) were obtained near 10 R_0, and curve (f) was obtained near 20 R_0. The straight dashed lines represent estimates of the shape and level of $D(s)$ at 10, 20 and 50 R_0 based on angular broadening measurements [3,4,5,6]. The triangles give typical structure function levels at 10, 20 and 50 R_0 as inferred from intensity scintillations [3,4]. The data points labelled (g) are $D(s)$ estimates from measurements of fluctuations in VLBI differential phase over baseline s [3,7]. These values were estimated directly from equation (2) based on the reasonable assumption that at these scales the observed phase is a good approximation to the geometric phase [3]. Finally, the solid straight lines in Figure 1 are estimates of D(s) at large scales based on spacecraft-beacon phase scintillations. Here we have used Woo and Armstrong's measurements of the phase scintillation power $W_\phi(f)$ at f=0.001 Hz to estimate the level of D(s) at large scales ($\sim 10^5$ km) [1,3]. This was then extrapolated to smaller scales assuming a Kolmogorov power law (α=11/3) at 10, 20 and 50 R_o and α=3.4 at 5 R_o.

The data in Figure 1 suggest that the microscale density spectrum near the Sun departs from a simple power law in a complex but consistent fashion. The D(s) $\propto s^2$ behavior of the angular broadening data and the breaks (steepening at small s) seen in the spectral broadening structure functions are consistent with the existence of an inner scale s_i which increases with increasing radial distance R. This is consistent with the size and R-dependence of the inner scale deduced from earlier intensity scintillation observations in the region R=12-70 R_o [8]. Above the inner scale the D(s) flatten to α values slightly less than 3.0, and it is this flat region (s=2-100 km) which dominates the S-band spectral broadening measurements for R>5 R_o [1,2,3]. The structure functions must steepen again at s>100 km in order to be consistent with the observed levels of phase scintillation power at large (10^5 km) scales.

In Figure 2 we have constructed model density spectra $\Phi_{ne}(\kappa)$ at three radial distances based on the results shown in Figure 1. Here we have idealized the spectrum as a two-component power law with a gaussian inner scale of the form $\exp(-\kappa^2/\kappa_i^2)$. For the power law

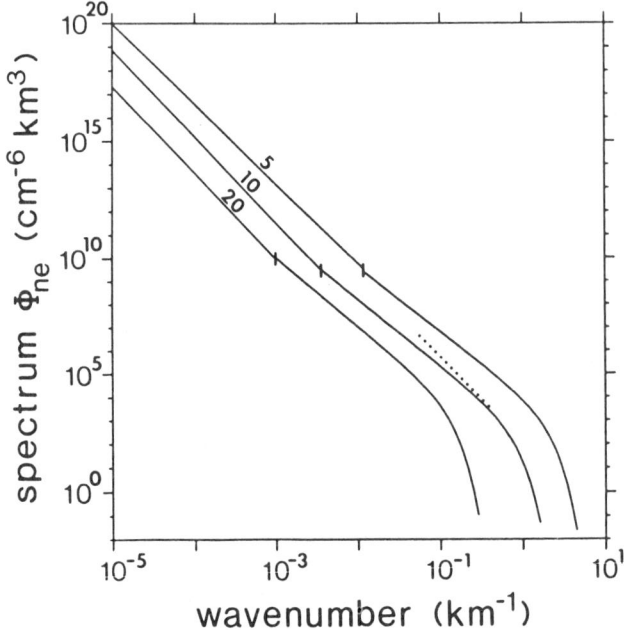

Fig. 2. Model density spectrum $\Phi_{ne}(\kappa)$ at 5, 10 and 20 R_o.

break (denoted by the vertical tics in Figure 2) we used $\kappa_b = 1/s_b$, where s_b was defined as the intersection of the phase scintillation D(s) with the extrapolation (denoted by a dotted line in Figure 1) of the spectral broadening D(s). For more details on the conversion from D(s) to $\Phi_{ne}(\kappa)$ the reader is referred to Coles and Harmon [3]. Also shown in Figure 2 is a dotted line representing the transient spectral steepenings seen in some of the Arecibo radar spectral broadening observations made in 1979 [2,3].

We conclude that the various radio propagation results for the near-Sun region (5-50 R_o) are best explained by invoking a relatively stable (but evolving) density spectrum whose shape departs from a simple power law in a systematic fashion at high spatial wavenumbers (small scales). The basic features of this model are: 1) a relatively steep (Kolmogorov) power law slope at large scales, 2) a local flattening at intermediate scales, and 3) a steepening at small scales suggestive of an inner scale. The work of Woo and Armstrong [1] suggests that the spectrum undergoes an overall flattening at large scales for $R \cong 2 R_o$, and our results suggest that this effect may start to be evident at 5 R_o. However, it appears that outside of 5 R_o the dominant characteristic of the spectrum is not a variation in overall slope (α) with distance R but rather variation in slope with spatial wavenumber.

Because our spectral model is based to some extent on comparisons of observations made at different times, additional work is needed to confirm its basic features. The most promising first step is the reanalysis of existing spacecraft propagation data in order to obtain simultaneous spectral broadening and phase scintillation spectra from the same data set. We are currently undertaking such an analysis using Voyager radio data; preliminary results indicate that our putative spectral break κ_b is a real and well-defined feature.

REFERENCES

1. R. Woo and J. W. Armstrong, J. Geophys. Res. 84, 7288 (1979).
2. J. K. Harmon and W. A. Coles, Ap. J. 270, 748 (1983).
3. W. A. Coles and J. K. Harmon, Ap. J. (in press).
4. W. A. Coles, Space Sci. Rev. 21, 411 (1978).
5. R. G. Blessing and P. A. Dennison, Proc. A.S.A. 2, 84 (1972).
6. W. C. Erickson, Ap. J. 139, 1290 (1969).
7. D. B. Shaffer, J. E. Salah, and A. E. E. Rogers, paper presented at January 1986 USNC/URSI Meeting, Boulder, Colorado.
8. S. L. Scott, W. A. Coles, and G. Bourgois, Astron. Astrophys. 123, 207 (1983).

SPATIAL COHERENCE IN A SCATTER-BROADENED IMAGE – OBSERVATIONS OF 3C279 CLOSE TO THE SUN

T.J. Cornwell, NRAO[†], Socorro NM 87801
K.R. Anantharamaiah[‡], NRAO, Socorro NM 87801
R. Narayan, Steward Observatory, Univ. of Arizona, Tucson AZ 85721

ABSTRACT

Using the NRAO Very Large Array radio-telescope, we have imaged the IPS scattered disk of 3C279 close to the Sun with high spatial and temporal resolution. The resulting images have speckles, similar to those seen in short-exposure optical photographs of the seeing disk. We describe a test for coherence in the scattered radiation, and confirm that, under suitable observing conditions, the observed radiation is indeed spatially coherent.

INTRODUCTION

For times short compared to the diffractive scintillation time-scale the scattering process in Inter-Planetary Scintillation (IPS) or Inter-Stellar Scintillation (ISS) is frozen, and, in many respects, it is similar to the scattering which occurs in a single-dish radio-telescope possessing significant aberrations. Recent work by Cornwell and Napier[1] has explored the attributes of the coherence function of the scattered radiation field. They showed that the coherence function contains information about the scattering screen, and that, given complete sampling of the coherence function in the focal plane of a radio-telescope, both the scattering screen and intrinsic source structure can be obtained. Therefore, drawing an analogy between the focal plane of a radio-telescope and the Earth's surface in IPS or ISS, we can use an existing radio-interferometric array, such as the NRAO Very Large Array (VLA)[2], to determine properties of both the scattering screen and the background object.

In strong scintillation theory, freezing the scattering screen corresponds to observing for a time short compared to the diffractive time-scale t_{diff}. This is difficult in ISS because t_{diff} ($\sim 10^2 - 10^3$ sec) is considerably shorter than the optimum imaging time for VLBI arrays (~ 1 day). However, in IPS at a solar elongation of a few degrees, $t_{diff} \sim 0.1$ sec at centimetric wavelengths and the VLA can provide high-quality, good SNR images of bright sources even with such short integration times. Further, the corresponding scattering disk, $\theta_s \sim 1-5$ arcseconds, is resolvable with the VLA, making this an ideal telescope for the study of an IPS scattered image.

[†]The National Radio Astronomy Observatory (NRAO) is operated by Associated Universities, Inc., under contract with the National Science Foundation.
[‡]On leave from Raman Research Institute, Bangalore, India.

© 1988 American Institute of Physics

In this paper we report VLA observations of the bright extragalactic radio source 3C279 as it passed close to the Sun. We will discuss aspects of images formed in all three time-integration regimes: $t \ll t_{diff}$, $t_{diff} < t < t_{ref}$, and $t \gg t_{ref}$, where t_{ref} refers to the refractive timescale of strong scintillation. We will call these regimes *diffractive, refractive* and *super-refractive*. (Goodman and Narayan, this volume, have also investigated the same regimes through computer simulations; they refer to the corresponding images as *snapshot, average* and *ensemble-average*, respectively.) In our observations, $t_{diff} \sim 0.1$ sec, and $t_{ref} >$ 10 sec.

THE OBSERVATIONS

The VLA is a 27-element radio-interferometric array located in the Plains of San Augustin, New Mexico[2]. The elements are spread over a Y-shape whose maximum separation can be chosen to be 1, 3, 10, or 35 km. Allowed observing wavelengths are 400, 90, 21-18, 6, 2, and 1.3 cm. Two independently tuneable data-paths exist. The on-line computer system limits the data-rate so that a complete set of correlation coefficients can be saved every $t_{data} \geq 6.6$ sec. However, it is possible to gate the correlator so that any integration time, t_{int}, down to 92 μsec is possible.

In our observations, the optimum value for the integration time, t_{int}, is determined by a compromise between SNR and smearing of the scintillation pattern. To resolve 4 points across the scattered image, we require $t_{int} < 40$ msec, independent of observing wavelength. For 3C279, the resulting SNR ranged from 10 to 100.

At sub-arcsecond resolution, self-calibration must be used to correct for the effects of the neutral atmosphere on the measured visibility phases[2,3]. To take advantage of this capability, we tuned both data-paths to the same sky-frequency. One was gated to obtain short integration visibility data (t_{int} = 10 and 40 msec), while the other was used, un-gated, for long integration visibility data (t_{int} = 6.6 sec). Since the long integration images were expected to be stable in structure, we planned to self-calibrate the long-integration data, and then transfer the resulting antenna phases to the short-integration data. In practice, this was necessary at wavelengths of 6 cm and 2 cm. Exploiting the simple Gaussian-like shape of the super-refractive images, we used only a simple point source model for the self-calibration step.

The calibrated visibility data were transformed into images using standard procedures[2], and the resulting images were deconvolved using CLEAN[4].

SNAPSHOT IMAGES

An example of a snapshot (or diffractive) image is shown in Figure 1a. Interpretation of these images has some pitfalls: it is important to remember that they have been constructed according to the theory appropriate for *spatially-incoherent* objects (see e.g. Thompson et al.[3]), whereas within a sufficiently small bandwidth and integration time the scattered radiation is spatially coherent. To understand the effect of this spatial coherence, consider an example in which the dominant scattering comes from just two spots on the screen. If the background object is sufficiently unresolved, and if both the bandwidth and integration time are sufficiently small, then the radiation from these two spots will be completely coherent. Far from the scattering screen, spatially-varying fringes will occur from interference of the radiation from these two spots. Hence in this example, and in the general case, if we measure the coherence function with an interferometric array, the main effect can be thought of as a modulation of the complex gains of the array elements. Thus for an N-element array, the $N(N-1)/2$ measured visibilities can be factorized into N complex numbers, the antenna gains. To verify this, we performed a least-squares fit to obtain the antenna gains, and we then made images of the corrected visibility. In the jargon of radio-interferometry, this corresponds to self-calibrating the antenna complex gain using a point source model, and so standard AIPS routines can be used. An equivalent test is to check that the closure phases are zero and that the closure amplitudes are unity. An example of a resulting self-calibrated image is shown in Figure 1b. The fact that this image resembles a single point source so well with no extraneous structure above the noise level proves that the received radiation is indeed perfectly spatially-coherent, thus confirming the prediction that we were testing.

The gain modulation on the antennas discussed above leads to rather subtle effects if conventional methods are used to process the image. This is because the measured coherence function is a function of absolute position of the elements, rather than of just the relative position as would be the case for an incoherent object. Since standard image processing methods assume an incoherent source, the samples are transformed according to relative position only. In the present case, adjacent Fourier components in the final image may have been measured by completely distinct parts of the aperture, with quite different scintillation effects. In the image plane, this causes a pseudo-scattering on a scale-size of about $\theta_{ps} \sim \lambda/\Delta r_{min}$, where Δr_{min} is the minimum increment in baselines. This is usually much larger than the true scattering angle $\theta_s \sim \lambda/l_0$, where l_0 is the diffractive scale-size. In a filled aperture, this effect is washed out by the large amount of redundancy and by the continuity of the aperture, but in an un-filled aperture, such as the VLA, the non-redundancy is very harmful.

Another aspect of the gain modulation is that the diffractive images show fine-scale structure not visible in the super-refractive image. By analogy to the bright spots seen in short-exposure photographs at the image plane of an optical telescope, we call these "speckles". Since the scintillation pattern has scale-size l_0 on the ground, the number of speckles in the image is $\sim (B/l_0)^2$, where B is the size of the array. Therefore, l_0 plays a role analogous to the Fried parameter in optical speckle. However, this parallel holds only to a limited degree: unlike their optical counterparts, these speckles do not consist of diffraction-limited copies of the background object. Instead, structure shows up in the observed visibility function directly, and so the coherence function on the ground is analogous to the coherence function (of which the diagonal part is the image) at the focal plane of an optical telescope[1].

As well as imaging the scattering disk, we have also attempted to image the scattering *screen* directly by transforming the derived antenna gain pattern. Unfortunately, samples of the gain pattern are obtained only at the positions of the array elements, and so the phase screen is very poorly constrained. For this to work properly, one would need an array in which the antenna *positions*, not just their separations, spanned a two-dimensional area.

REFRACTIVE AND SUPER-REFRACTIVE IMAGES

Although basically Gaussian-like in shape, the refractive images show a large amount of variation. For a number of the data sets, we are investigating the relationship between the total flux of a refractive image and the size of the scattering disk. When a sufficient number of independent diffractive images are averaged (by using large integration time and bandwidth), the flux and the major axis appear to be well-correlated, as predicted[5].

Super-refractive images were made by taking all the long-integration visibility data for the whole observing time, typically several to 10 minutes (Figure 1c). The resulting images were smooth but highly anisotropic, consistent with the results of Armstrong et al.[6]. The major axis is almost tangential with respect to the Sun.

DISCUSSION

We have investigated the properties of snapshot images of the scattering disk in IPS. We have demonstrated a simple test for the coherence of the scattered radiation: if the measured coherence function factorizes into parts dependent purely upon antenna position then the scattered radiation is completely coherent. A failure of this test when the integration time and bandwidth are sufficiently small would reveal lack of coherence in the electric field before scattering, which in turn would indicate the presence of resolved structure in the background object. We are now pursuing the possibility of using such information for high-spatial-resolution imaging in both IPS and ISS.

Although in this paper we have concentrated upon the optics of our experiment, the observations also contain interesting information about the solar wind. A more systematic investigation of this aspect is underway.

Acknowledgements: We thank K. Sowinski for suggesting the correlator-gating technique used in the VLA observations. Discussions with P.J. Napier and J. Goodman are gratefully acknowledged. This work was supported in part by NSF grant AST-8611121.

REFERENCES

1. T.J. Cornwell, and P.J. Napier, "Self-correction of Telescope Surface Errors Using a Correlating Focal Plane Array", Proc. NRAO Greenbank Workshop on "Radio Astronomy from Space", ed. K. Weiler, (1986).
2. P.J. Napier, A.R. Thompson, and R.D. Ekers, *Proc I.E.E.E.*, 71, 1295–1320, (1983).
3. A.R. Thompson, J.M. Moran, and G.W. Swenson, *Interferometry and Synthesis in Radio Astronomy*, Wiley Interscience, (1986).
4. J. Hogbom, *Astrophys. J. Suppl.*, 15, 417–426, (1974).
5. R.Blandford, and R.Narayan. *Mon. Not. R. astr. Soc.*, 213, 591-611, (1985).
6. J.W. Armstrong, W.A. Coles, M. Kojima, and B.J. Rickett. *The Sun and the Heliosphere in Three Dimensions*, pages 59–64, ed. R.G. Marsden, D. Reidel, (1986).

Figure 1: Images of 3C279. Solar elongation = 0.9 degrees, $\lambda = 2$ cm, bandwidth = 50 MHz. Contour levels: -90,-80,...,-10,-5,5,10,...90 % of peak brightness. Tick marks at intervals of 0.2 arcsec. (a) Diffractive Image, $t_{int} = 40$ msec. (b) Self-Calibrated Diffractive Image. (c) Super-Refractive Image, $t_{int} = 6$ min.

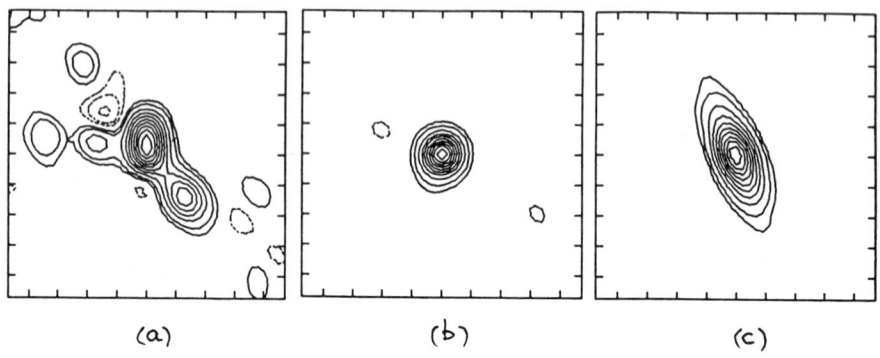

INTERFEROMETRIC PHASE SCINTILLATION IN THE IPM

B. Dennison (VPI & NRL), R. S. Simon (NRL),
S. Ananthakrishnan (Ooty), and R. L. Fiedler (NRL)

ABSTRACT

Using a global VLBI array operating at 327 MHz, we find strongly correlated scintillations in the fringe phase obtained on various baselines. The magnitude of the fringe phase scintillations is approximately proportional to baseline length. Cross correlation of the fringe phase on different baselines reveals maximum correlation at some small lag, with the value dependent upon baseline pair. We find that the lags of maximum correlation (for different baseline pairs) are fit very well by a model in which the phase fluctuation pattern is moving across the array with fixed velocity. Since the fit velocity is 410 km/s almost radially outwards from the sun, we confirm that the interplanetary medium is responsible. The implied phase power spectrum is in good agreement with other observations.

These results have important implications for VLBI, particularly at low frequencies.

INTRODUCTION

For quite some time it has been established that density irregularities in the solar wind can, under certain conditions, cause scintillations in the flux received from distant, compact radio sources (e.g. Hewish et al. 1964). The phase fluctuations in the emergent wavefront (which diffractively result in intensity scintillations) can, of course, be detected directly. This has been exploited in space probe experiments in which the phase of the received carrier wave has been monitored (Woo and Armstrong 1979).

Here, we present preliminary results from observations of scintillations in the fringe phase of a VLBI array. This technique is a very sensitive probe of density irregularities in the solar wind, because it does not require that measurable intensity scintillations be present. In comparison with space probe experiments, the frequencies available for such VLBI experiments are generally lower than those used on space probe uplink/downlinks, resulting in much higher sensitivity to fluctuations on scales comparable to and smaller than the baselines used.

© 1988 American Institute of Physics

OBSERVATIONS AND RESULTS

Phase scintillation was noticed on all baselines in a global VLBI experiment at 327 MHz, the primary purpose of which was to study extragalactic sources. Figure 1 shows the station-based phases, produced by global fringe-fitting, for the calibration source, 3C 286, during a 30-minute period on December 10, 1983, when the solar elongation angle was 75 degrees. In producing these phases, the phase of the reference antenna, the NRAO 140-foot (NRAO), was held to zero. Significantly, the magnitude of the fluctuations increases with increasing baseline in a manner which is in accord with that expected from a moving, phase-changing screen with an approximately Kolmogorof spectrum of turbulence.

As a model to explain the phase fluctuations shown in Figure 1, we consider a screen which moves across the array at constant velocity. The phase fluctuations within the screen are assumed to be frozen-in on timescales less than or equal to the array-crossing time.

With this model in mind, the fringe phases (referenced to the NRAO 140') on the baselines NRAO-X were cross-correlated with a reference baseline, NRAO-Arecibo. One such cross-correlation is shown in Figure 2. The lag of peak cross-correlation, τ, gives the delay in the fluctuations between the two baselines. The values of τ, thusly determined for various baseline pairs, were fit to the relation,

$$\tau = \frac{\vec{D} \cdot \hat{n}}{|\vec{v}|} = \frac{\vec{D} \cdot \vec{v}}{|\vec{v}|^2} ,$$

where $\vec{D}$ is the difference vector connecting the midpoints of the two baselines in a pair, $\vec{v}$ is the velocity of the phase scintillating screen, and $\hat{n}$ is a unit vector in the direction of $\vec{v}$. All of the vectors are projections in the sky plane. The difference vectors are shown in Figure 3a. Least-squares solution to the above equation yields $|\vec{v}|$ = 410 km s^{-1}, at a position angle = -43 degrees. As this is nearly radially outward from the sun, we confirm that the solar wind is responsible (Figure 3b). In Figure 4, we show the measured lags versus those predicted by the least squares solution. Clearly the fit is quite good.

From the magnitude of the observed fluctuations, we have computed the phase power spectrum in the IPM, and find it to be consistent with ground-based studies of intensity scintillation (e.g. Gapper and Hewish 1981), as well as space probe studies of phase scintillation (Woo and Armstrong 1979). The assumed "normal" strength of

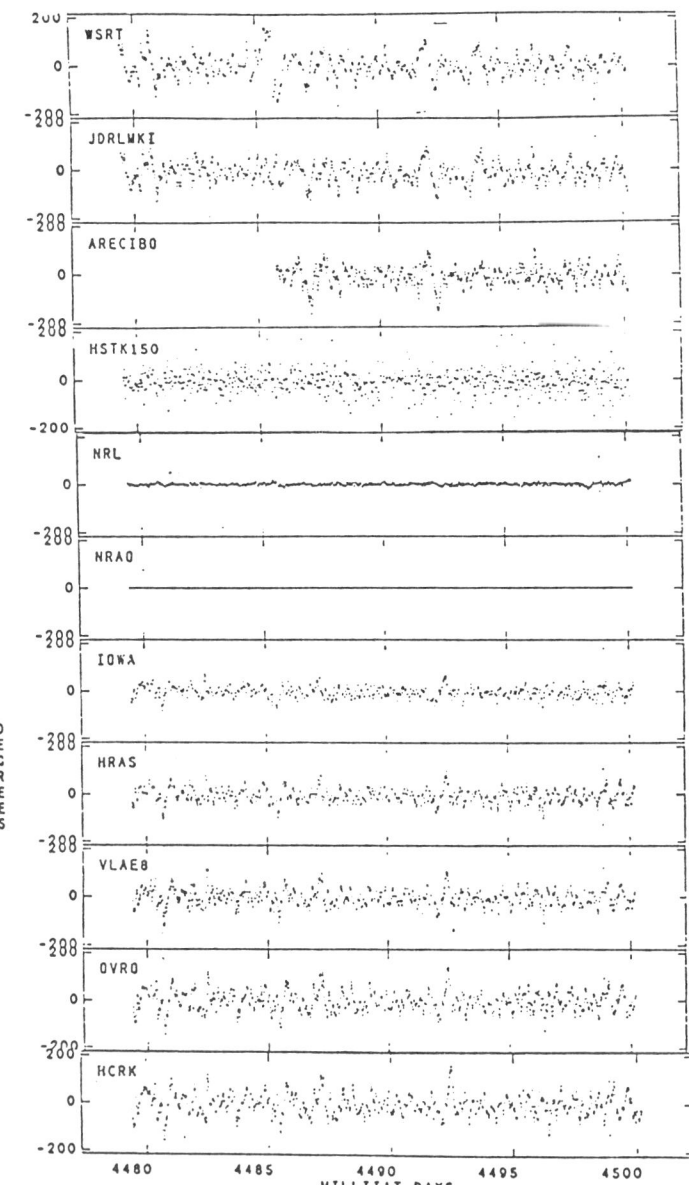

Fig. 1. Station-based phases referenced to the NRAO 140-foot. Any long term fluctuations have been removed by the fitting process. These phases cover a time span of 30 minutes. The source (3C 286) was at a solar elongation = 75 degrees.

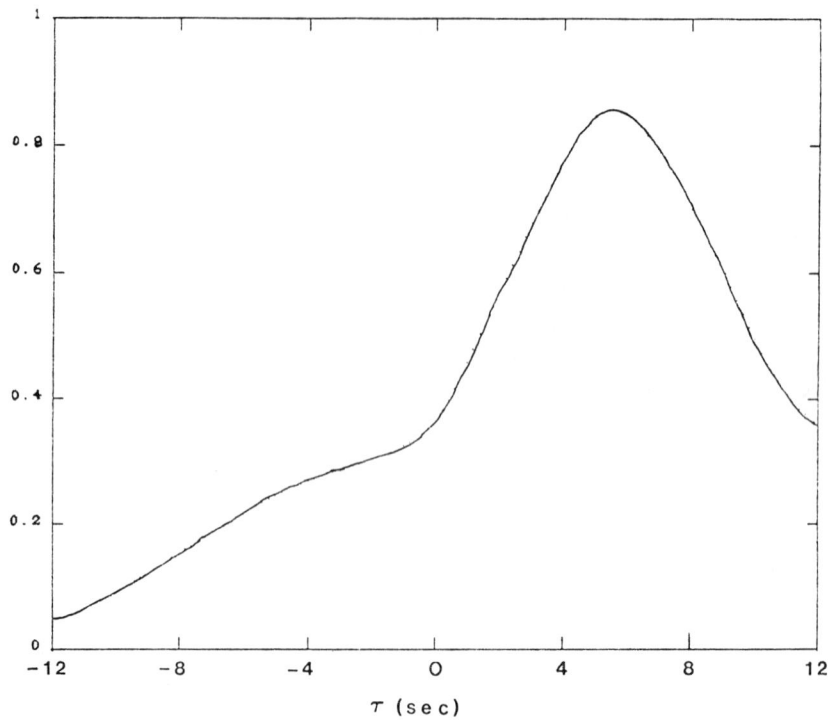

Fig. 2. Cross-correlation of the fringe phase on the NRAO-OVRO baseline with the reference baseline, NRAO-Arecibo. Note the strong maximum cross-correlation at non-zero lag, indicating a delay in the fringe phase fluctuations between these two baselines.

Solar Wind Turbulence

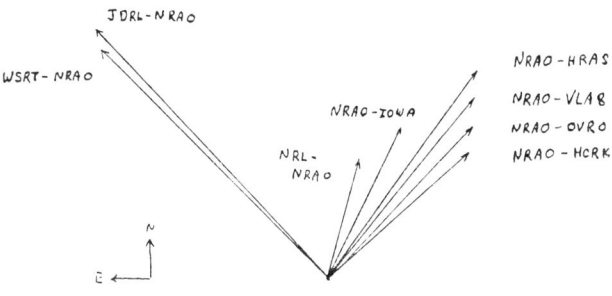

Fig. 3a. The projected difference vectors between the baselines indicated and the reference baseline, NRAO-Arecibo. The diagram is in the sky plane as seen by an observer.

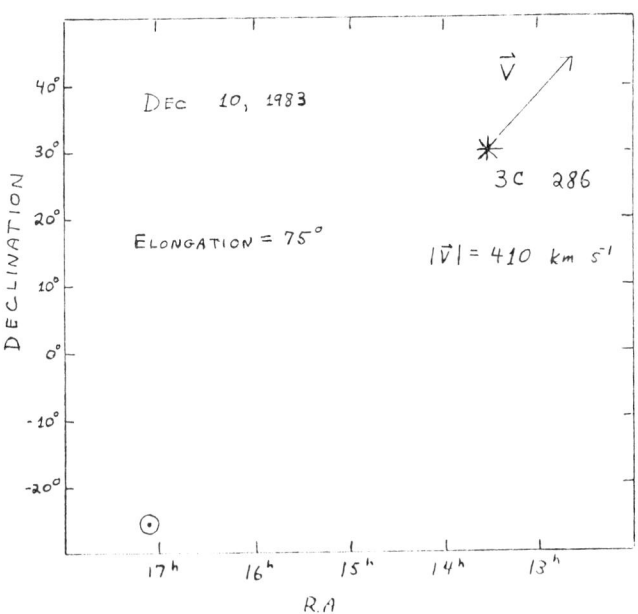

Fig. 3b. The positions of 3C 286 and the sun at the time of observations. The vector velocity of the flow responsible for the phase scintillations is shown.

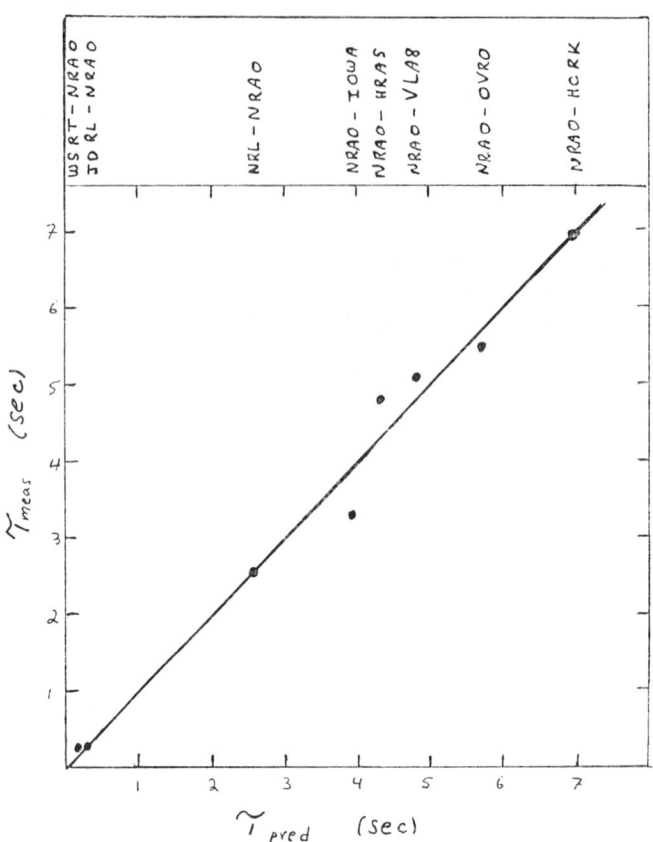

Fig. 4. Measured lags of peak cross-correlation versus those predicted by the least-squares solution. Indicated across the top are the baselines which were cross-correlated with the reference baseline, NRAO-Arecibo.

interplanetary turbulence can readily account for the fringe phase scintillation reported here.

DISCUSSION

This technique provides a very sensitive probe of density fluctuations in the IPM. It could therefore be exploited to study solar wind fluctuations in regimes not normally accessed by observations of intensity scintillations. Because of the lower frequencies available for VLBI, this technique may also be more sensitive than measurements of phase scintillation using spacecraft links.

Measurements of this type directly probe the phase structure function of the medium, as effectively given by the rms fluctuation in the fringe phase as a function of projected baseline. We have not exploited this aspect of the experiment because long time scale fluctuations ($\geq$ 30 seconds) were filtered in some poorly understood way by the fringe fitting process. VLBI finges can be analyzed in a manner that avoids this problem, however.

Our results have important implications for VLBI and low-frequency interferometry. In Table I we summarize the magnitude of the effects expected for various arrays. τ is the decorrelation timescale and $\bar{\phi}_{RMS}$ is the rms phase fluctuation. Clearly VLBI at frequencies of hundreds of MHz, and on baselines of thousands of kilometers, will suffer significant decorrelation when integration times exceed several seconds. With the extension of baselines to satellite-based antennas the magnitude of the fluctuations will be larger. The decorrelation can be avoided if sufficiently short integration times are used, in which case it may be possible to remove the phase fluctuations using station-based fringe fitting techniques. Of course, these effects are somewhat smaller at larger solar elongations. At the lowest frequencies (several MHz) envisioned for the Low-Frequency Space Array (LFSA) the interplanetary scattering will be strong. In this case, the effective image broadening in the anti-solar hemisphere will be comparable to the limitations already set by interstellar scattering.

TABLE I - Decorrelation Caused by the IPM

Array	Frequency (MHz)	Baseline (km)	Φ_{RMS} (rad)	τ (sec)
VLBI	327	1000	0.5	-
VLBI	327	10000	3.4	5
Quasat	327	20000	6	4
Radio-Astron	327	70000	17	4
LBI	81	1000	2	1
LFSA	2	10	1	0

REFERENCES

Gapper, G. R., and Hewish, A. 1981, M.N.R.A.S., 197, 209.

Hewish A., Scott, P. F., and Wills, 1964, Nature, 203, 1214.

Woo, R., and Armstrong, J. W. 1979, J. Geophys. Res., 84, 7288.

IV. DIFFRACTION PHENOMENA AND THE INTERSTELLAR ELECTRON DENSITY POWER SPECTRUM

VLBI Observations of the Scattering Disk of Pulsar 1933+16

C.R. Gwinn (Center for Astrophysics), J.M. Cordes (Cornell U.), N. Bartel (Center for Astrophysics), A. Wolszczan (NAIC), R. Mutel (U. Iowa)

Abstract

A Kolmogorov spectrum of density fluctuations in the interstellar plasma is favored by VLBI observations of the scattering disk of pulsar 1933+16. The decline of visibility with baseline length and the scaling of size with frequency agree with those predicted on the basis of such a spectrum. Solving for the spectral index of density fluctuations, α, from the visibility as a function of baseline length, we obtain $\alpha = 3.52 \pm 0.13$. We place an upper limit of 1:1.7 on the elongation of the scattered image.

Introduction

Pulsars are excellent probes of interstellar scattering. They are pointlike at angular scales probed by VLBI, so that any observed structure must be due to interstellar scattering. Alone among known varieties of radio sources, they provide information about the interstellar plasma through pulse dispersion, pulse broadening, and interstellar scintillation. VLBI provides information complementary to these phenomena.

To study interstellar scattering, we have observed several pulsars with intercontinental VLBI networks. Pulsar 1933+16 was almost completely resolved on our longest baselines at a frequency of 326 MHz; its size was therefore well matched to our (u,v) coverage. Here we focus on observations of that pulsar's scattering disk, and their interpretation.

Observations

We observed 1933+16 with the US and European VLBI networks at frequencies of 326 and 608 MHz. We used the Mark II VLBI system with a 2 MHz bandwidth. The 326 MHz observations were made on 23 March 1986 with telescopes at Westerbork (phased array), Jodrell Bank (76 m), Arecibo (305 m), Haystack (46 m), Maryland Point (26 m), Green Bank (43 m), Fort Davis (26 m), and Owens Valley (40 m). The 608 MHz observations were made on 8 October 1986 with telescopes at Bonn (100 m), Westerbork (phased array), Jodrell Bank (76 m), Arecibo (305 m), Green Bank (43 m), Iowa (18 m), and Owens Valley (40 m). We recorded left-circular polarization (IEEE convention) at all stations with the exception of Arecibo, which can observe only linear polarization at these frequencies due to feed limitations. During the observations, pulsar pulse phase and period information was recorded onto the cassette at the Green Bank telescope. The NRAO Mark II correlator in Charlottesville, Virginia, used this information to inhibit correlation off-pulse, yielding a factor of 2 to 5 increase in the signal-to-noise ratio (Gwinn et al. 1986).

The (u,v) coverage for pulsar 1933+16 was elongated and would have been almost linear, with a position angle of 80° east of north, except for baselines to Arecibo. Information on the 2-dimensional structure of the source, including elongation, therefore depends on accurate calibration of the Arecibo gain.

Our observations were averaged over the diffractive timescale and bandwidth, τ_{ISS} and $\Delta\nu_{ISS}$, but not over the refractive timescale τ_{REF}, as Table

I indicates. For the model fits described below, the data were coherently integrated for 32 sec; integrations for 16 to 64 sec yielded results consistent with these. The coherence time, set by ionospheric variations at our observing frequencies, appears to be of the order of 128 sec.

Table I: Scattering Parameters for Pulsar 1933+16[a]

Parameter	Units	Symbol	326 MHz	608 MHz
Decorrelation Bandwidth	kHz	$\Delta \nu_{ISS}$	0.1	1.8
Diffraction Pattern Scale	cm	S_{ISS}	10^8	2×10^8
Scintillation Time	sec	τ_{ISS}	8	18
Refractive Timescale	yr	τ_{REF}	1.6	0.4
Strength of Scattering	$m^{-20/3}$	C_n^2	$10^{-2.63}$	$10^{-2.63}$

[a] From Cordes (1986).

The linear polarization recorded at Arecibo presented several problems. Since many pulsars, including 1933+16, have significant linear polarization, the fraction of signal recorded there could vary with the parallactic angle of the feed and ionospheric Faraday rotation, effectively causing gain variations. Moreover, polarization impurities of antenna feeds can lead to non-zero closure phases, even for pointlike sources (Bartel et al. 1985). Arecibo presents the worst case, since the unwanted right-circular polarization is of about the same strength as the desired left-circular polarization there, and has the maximum effect on any, presumably small, polarization defects at other stations. At 326 MHz, closure phases for all triplets of stations, including those involving Arecibo, were consistent with zero. At 608 MHz, closure phases were consistent with zero for all triplets except those involving Arecibo. Some of these closure phases showed large deviations from zero, with time variations which tracked those of the parallactic angle of the feed at Arecibo closely. We therefore deleted all 608 MHz data from baselines including Arecibo.

Data Calibration and Reduction

To treat telescope gains and source parameters simultaneously and on an equal footing, and to adjust for pulsars' large intrinsic flux density variations from pulse to pulse, we wrote a special least-square fit self-calibration program. This program fits a model to the correlated flux densities. The model can include gains at the stations and the size and index of the scattering disk as global parameters, and the flux density of the pulsar, the gain at Arecibo, and the station phases, as local parameters, which vary independently for each integration time.

The model correlation coefficient, $C_{AB}(t)$, of signals from stations A and B at time t is given by

$$C_{AB}(t) = \Gamma_A \Gamma_B S(t) \exp\left[-\frac{1}{2}\left(\frac{\pi}{\sqrt{2\ln 2}}\theta_H B_{AB}\right)^{\alpha-2}\right] \exp\left[\phi_A(t) - \phi_B(t)\right]$$

Here, Γ_A and Γ_B are the gains of the stations. The flux density of the pulsar is $S(t)$. The parameter θ_H is a measure of the scattering disk size, and is the

FWHM of the disk for $\alpha = 4$. Departure of the scattered image from a Gaussian, as predicted by power-law spectra of density irregularities, is parameterized by α. Elongation of the scattering disk is modeled by the axial ratio ϵ and position angle of the major axis, ρ. These parameters scale u and v, the traditional projected components of the interferometer baseline in wavelengths:

$$B_{AB} = \left| \begin{pmatrix} \epsilon \cos \rho & -\epsilon \sin \rho \\ \sin \rho & \cos \rho \end{pmatrix} \begin{pmatrix} u \\ v \end{pmatrix} \right|$$

The parameters $\phi_A(t)$ and $\phi_B(t)$ model the time-dependent phase changes at each station, as introduced by astrometric errors and changes in ionospheric path length.

The technique of sequential least squares (e.g. Kaula 1966) reduces the order of the matrices of partials from the total number of parameters, as for a standard least squares fit, to the number of global and local parameters for one time interval. It thus increases the speed and accuracy of the fitting. By this technique, the global parameters are sequentially updated using data from each time interval; their values and variance-covariance matrix are then used to find the local parameters. Since our model is nonlinear, the fit must be iterated. We augment the matrices of partials at each iteration to limit the changes in the parameters, until the fit has converged.

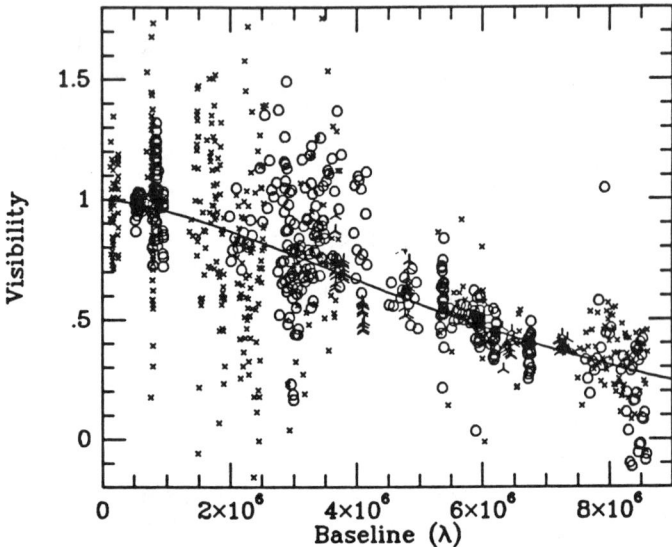

Fig. 1. Visibility as a function of baseline length, after self-calibration, for pulsar 1933+16 at 326 MHz. The data have been integrated for 160 sec after fitting, to reduce scatter due to the signal-to-noise ratio, and projected onto the time-dependent axis defined by the station phases ϕ_A and ϕ_B. Skeletal triangles indicate baselines involving Arecibo. Crosses indicate baselines involving Maryland Point or Ft. Davis, the smallest antennas. Circles indicate other baselines. Baseline lengths have been adjusted to remove effects of the elongation of the source, a small correction except for baselines involving Arecibo. The solid line shows the best-fitting model, as summarized in Table II.

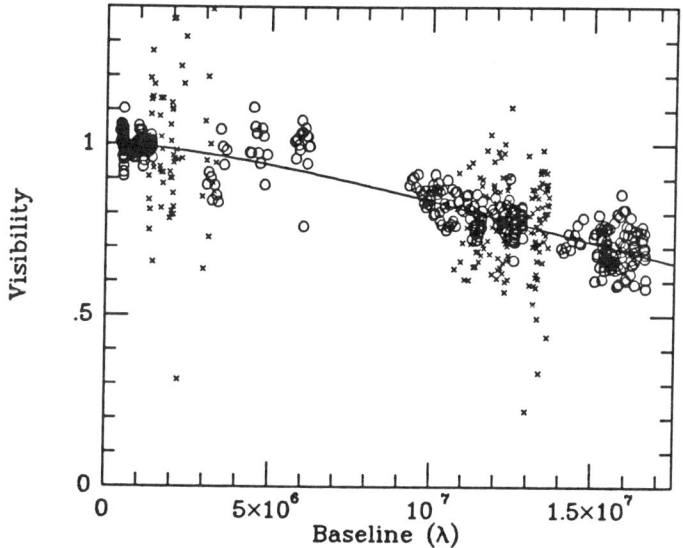

Fig. 2. Visibility as a function of baseline length, after self-calibration, for pulsar 1933+16 at 608 MHz. The data have been integrated for 160 sec after fitting, to reduce scatter due to the signal-to-noise ratio, and projected onto the axis defined by the station phases ϕ_A and ϕ_B. Crosses indicate baselines involving Iowa, the smallest antenna. Circles indicate other baselines. The solid line shows the best-fitting model, as summarized in Table II.

Results and Discussion

The model we fit to the amplitudes and phases of the correlated flux density of 1933+16 at 326 MHz included parameters θ_H, α, ϵ and ρ as well as station gains and phases. Figure 1 shows the visibility as a function of baseline length after correction for flux density, station gains, and the elongation of the image, as determined by the model fit. Table II summarizes the best-fit parameters. The stated errors are determined from the residual errors of the fit. The size is close to that of 20 mas estimated from previous measurements of decorrelation bandwidth and pulse broadening (Cordes, Weisberg and Boriakoff 1985 and references therein). Our fitted index α for the image is consistent with the value of $\alpha = 11/3$ expected for a Kolmogorov spectrum. If the Arecibo data are deleted, parameters θ_H and α are consistent with those in Table II, with

Table II: Fitted Parameters for Pulsar 1933+16

Parameter	Units	Symbol	326 MHz	608 MHz
Size	mas	θ_H	16.4 ± 0.6	3.99 ± 0.14
Index		α	3.52 ± 0.13	3.52^a
Elongation		ϵ	1.55 ± 0.15	---
Position Angle	deg	ρ	-17 ± 10	---

[a] Held constant for fit.

about the same uncertainties, so their fitted values are not strongly influenced by the Arecibo gain.

The elongation ϵ and position angle ρ are very sensitive to the gain at Arecibo. The time-varying gain at Arecibo as found from the fit tracks the parallactic angle there closely, confirming that 1933+16 is linearly polarized and that our model for the correlation coefficient is appropriate. This linearly polarized radiation recorded at Arecibo could corrupt the fitted values of ϵ and ρ, by spurious correlation with impure polarizations at other antennas as discussed above, although the closure phases show no evidence of such impurities. We therefore regard the observed elongation as an upper limit of $\epsilon < 1.7$ at present.

At 608 MHz, the data were sufficient to determine only the scattering disk size. We discarded Arecibo data, as noted above, and were unable to fit for source elongation ϵ and position angle ρ. The baselines are too short, given the source size, to find both size θ_H and index α. The fitted value for θ_H depends on the value assumed for α. We took $\alpha = 3.52$, as given by the 326 MHz data, and found $\theta_H = 3.99 \pm 0.14$ mas. Figure 2 shows the self-calibrated visibilities. Including the uncertainty of α determined from our 326 MHz data as another possible source of error in θ_H at 608 MHz, we find $\theta_H = 3.79 \pm 0.31$ mas. Assuming that the size scales with frequency ν as $\theta_H \propto \nu^{-\gamma}$, we find $\gamma = 2.28 \pm 0.19$ from θ_H and α at 326 MHz and 608 MHz and their standard errors, consistent with either the Kolmogorov value of $\gamma = 11/5$, or our fitted value for α at 92 cm, with $\gamma = \frac{\alpha}{\alpha-2}$ for $\alpha < 4$.

Conclusions

For pulsar 1933+16, visibility as a function of baseline length favors a spectrum of density fluctuations with an index of about the Kolmogorov value, $\alpha = 11/3$. These observations probe the spectrum of density irregularities near the baseline lengths, at length scales of 10^8 to 10^9 cm. From the scaling of visibility with baseline length at 326 MHz, we find $\alpha = 3.52 \pm 0.13$. We place an upper limit of 1:1.7 on the elongation of the scattering disk. These results are consistent at the $1.5 - \sigma$ level with those of Cordes, Weisberg and Boriakoff (1985), who found that the decorrelation bandwidth $\Delta\nu_{ISS}$ scales with frequency ν as $\Delta\nu_{ISS} \propto \nu^{4.22 \pm 0.16}$ for 1933+16, yielding $\alpha = 3.80 \pm 0.13$.

REFERENCES

1. N. Bartel, M.I. Ratner, I.I. Shapiro, R.J. Capallo, A.E.E. Rogers and A.R. Whitney, *A. J.*, **90**, 2532 (1985).
2. J.M. Cordes, *Ap. J.*, **311**, 183 (1986).
3. J.M. Cordes, J.M. Weisberg and V. Boriakoff, *Ap. J.*, **288**, 221 (1985).
4. C.R. Gwinn, J.H. Taylor, J.M. Weisberg and L.A. Rawley, *A. J.*, **91**, 338 (1986).
5. Kaula W.M., *The Theory of Satellite Geodesy*, Blaisdell Publishing Co.: Waltham, Mass., p. 125 (1966).

INTERSTELLAR SCATTERING OF SgrA*

D. C. Backer

Astronomy Department and Radio Astronomy Laboratory
University of California, Berkeley 94720

ABSTRACT

SgrA* is a compact, nonthermal radio source located in the center of the dense stellar cluster that defines the center of our galaxy. The effects of interstellar propagation, or limits thereon, on signals from SgrA* will be described.

ANGULAR SIZE

The summary by Davies, Walsh and Booth (1976) of radio interferometer observations of SgrA* between 0.96 and 8.1 GHz indicated that the apparent size of this source varies with λ^2. Subsequent VLBI observations by Lo et al. (1981, 1985) confirmed this dependence up to 22.3 GHz. They found a relationship of $\lambda^{2.0 \pm 0.1}$. Figure 1 displays published measurements of the apparent size of SgrA* from these references along with a $\lambda^{2.0}$ line. The results near 8 GHz are notable since they are spread over 8 years and give no indication of change with time to a limit of 0.6 mas y^{-1} (Lo et al. 1985). The agreement of these angles with λ^2 suggests an interstellar scattering origin and leads to an assessment that the intrinsic size of the source must be less than 10 AU at 22.3 GHz.

The angular broadening of SgrA* is anomalous when compared with interstellar scattering along nearby paths through the galaxy. Reference sources used in a VLA astrometry experiment on SgrA* (Backer and Sramek 1987) are smaller although they are extragalactic and traverse twice the path across the galaxy when compared with SgrA*; these objects are near 0.5° which means that their galactic trajectory is above a Z-height of 50 pc for half the path. The molecular masers in SgrB2 have apparent sizes near 0.25 mas at 22.3 GHz, an order of magnitude smaller than that of SgrA* at the same frequency (Gwinn et al. 1988). These maser signals traverse similar low Z-height path lengths through the inner galaxy. On the other hand Spangler and Cordes (1988) present evidence for very strong scattering of extragalactic sources within 0.5° galactic latitude and near longitude 25°. Further studies of compact objects along the plane near the galactic center are required to ascertain the 'expected' scattering of SgrA*.

If the SgrA* scattering is anomalous, then the scattering medium is probably located within a few hundred parsecs of SgrA* (Backer 1978). The proximity of the scattering medium to the emission site makes the scattering even more anomalous since the scattering angle must overcome the diverging angle of the spherical wavefront. The apparent source size θ_o is the product of the scattering angle defined in this volume by Rickett as θ_s and the demagnification factor $D = l/L$, where l is the source-scatterer distance and L is the source-observer path length. The scattering angle θ_s is a fixed property of the scatterer independent of its location. If the demagnification is small, then the scattering angle, which is a measure of the level of turbulence in the scatterer, must be greater than the apparent size θ_o by D^{-1}.

© 1988 American Institute of Physics

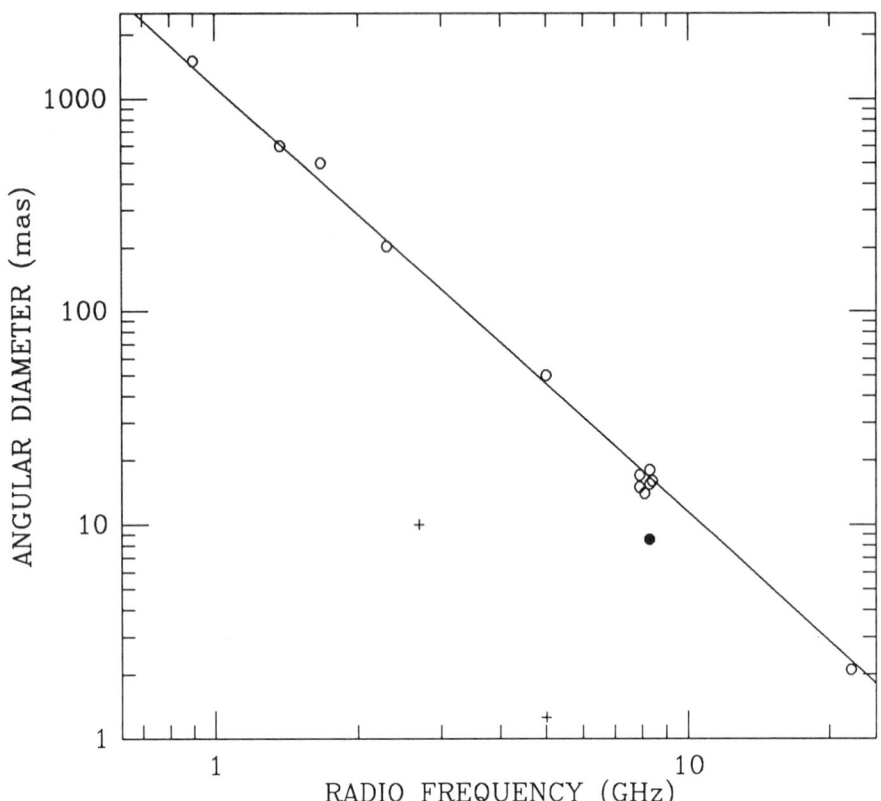

Fig. 1 Gaussian angular diameter (FWHM) determinations at various radio frequencies (open circles). Plus symbols are limits on refraction discussed in text. Filled circle indicates minor axis measured at 8.4 GHz.

Since the early assessments of the interstellar scattering of SgrA* two galactic paths have been found which show equally strong scattering: (1) a point source seen through the molecular cloud and ionized gas in NGC 6334 (Rodríguez et al. 1982); and (2) PSR 1849+00 seen through the periphery of a supernova remnant (Clifton 1986). In both cases the scattering medium is probably midway along the signal path through the galaxy. These occurrences of strong scattering point out that when such excesses exist they are related to an identifiable region of turbulence. The central few hundred parsecs of the galaxy is a reasonable site for the anomalous scatttering of SgrA*. There is a rotating molecular ring at a radius of 300 pc, and a second ring of neutral, molecular, and ionized gas with a radius of 1-10 pc. Mezger and Pauls (1978) identify an extended HII ellipsoid with a 150-pc by 50-pc extent in longitude and latitude, respectively, and an emission measure of 5×10^4 pc cm^{-6}.

We can estimate the level of turbulence in the SgrA* case from the observed scattering when suitable assumptions are made. Our starting point is the expression for the ensemble average visibility function for a source embedded in the turbulent medium. Coles *et al.* (1987) give this as

$$V(\sigma) = \exp\left[-2\pi r_e^2 \lambda^2 \int_0^L dz \int_\infty^\infty d\mathbf{k} \left[1 - \cos(\frac{\mathbf{k}\cdot\sigma\,z}{L})\right] P(k_x, k_y, k_z = 0; z)\right] \quad (1)$$

Equation 1 can be compared to Rickett's equation 3 in this volume. The presence of the demagnification factor in the cosine term of the power spectrum integral means that smaller length scales are relevant to the integral near the source. Eventually these length scales will fall below where the power spectrum is cutoff by dissipative processes, the inner scale. In the case of SgrA* at cm wavelengths with scattering within a few hundred parsecs of the center we are dealing with scales of $3\times 10^{4-5}$ cm; even if the scattering occurs midway to the source the scales are 10^{6-7} cm. In this case the decorrelation of the visibility function is probably the result of gradients from much larger scale turbulence. Coles *et al.* discuss the scattering for a power-law spectrum with index $\beta \equiv \alpha + 2 < 4$ and an exponential cutoff at an inner scale of k_0:

$$V(\sigma) = exp\left[-\pi^\alpha \Gamma(1 - \frac{\alpha}{2}) k_0^{2-\alpha} \left(\frac{r_e \lambda \sigma}{L}\right)^2 \int_0^L dz\, C_n^2(z)\, z^2\right] \quad (2)$$

Equation 2 tells us three things. First the coherence function is gaussian since it depends strictly on σ^2. This results from the expansion of the Fresnel filter for small wavenumbers where there is power in the turbulence spectrum. Nongaussian visibility functions arise only when the Fresnel filter interacts with the spectrum; *i.e.*, when the spectrum is nonzero on scales of the demagnified baseline $\sigma\,z/L$. Second the gaussian fits to the apparent size will strictly follow a λ^2 dependence as observed (Lo *et al.* 1985).

Finally we need an estimate of the inner scale and a location of the scattering to obtain the amplitude of the turbulence spectrum. For an inner scale of 10^8 m and a 200-pc thick scattering screen at $z = 200$pc, the value of C_n^2 is 5×10^6 m$^{6.67}$. The corresponding rms electron-density amplitude based on integrating a decade of the spectrum is 1.0 cm^{-3}. The detection of this source down to 1 GHz limits the column density of electrons in any screen by the free-free opacity. For the 200-pc screen used here, the density must be less than 100 cm^{-3} which is comfortably larger than the turbulence estimate. The density estimate can be reduced by decreasing the inner scale, but the effect is rather weak for power-law indices near 11/3.

The VLBI results of Lo *et al.* (1985) at 8 GHz indicated that the scattering profile was elliptical. This important result requires the scattering, and therefore the turbulence spectrum, to be anisotropic. The data from this experiment have been replotted by Marr (1988) as log[ln[V]] against log[σ] to display the form of the visibility (Fig. 2). The data from the GSTN-HCRO baseline at baselines near $10^{6.7}\lambda$ are the principal evidence for ellipticity since they are at postion angles near 0° in contrast to most of the data which have position angles between 30° and 45° . The slope of the line plotted in Figure 2 is 2.2 which is unexpected based on the model presented above. We are investigating the dependence of this slope on the zero baseline flux density and angular size, and plan to repeat the observations to obtain more *uv*-coverage.

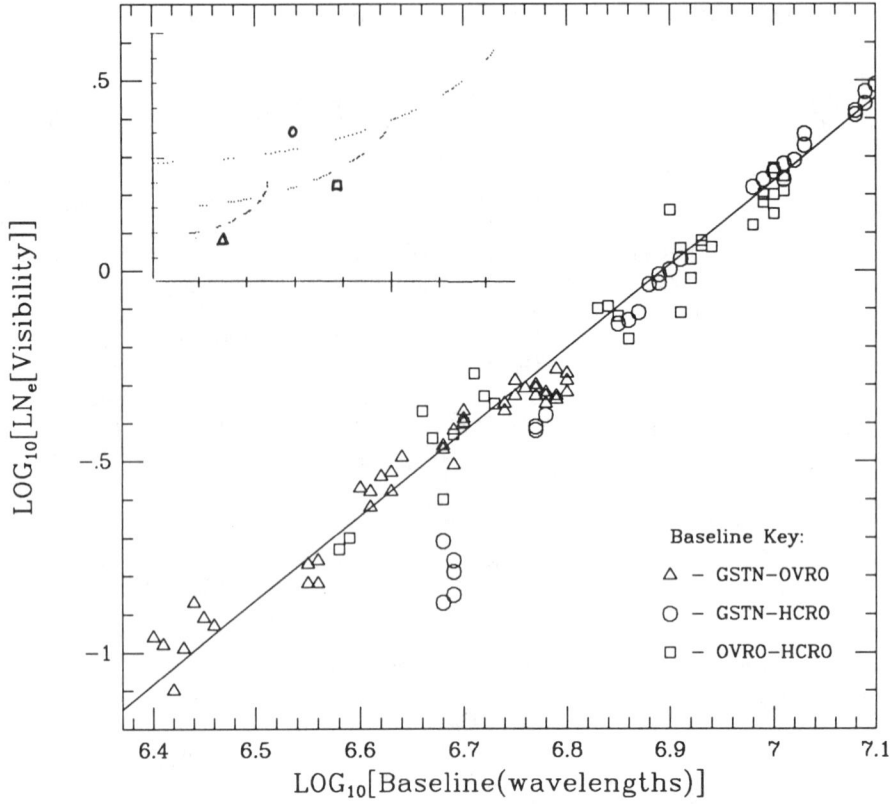

Fig. 2 Visibility of SgrA* from the 1983.4 VLBI observations presented by Lo *et al.* 1985) presented to display the power-law dependence of the natural logarithm on the baseline. Total flux density used was 1.26 Jy. The projected baselines are shown in the inset. The slope of the line drawn is 2.2.

PROPER MOTION

In 1974 Sramek and I began measurements to determine the proper motion of SgrA* (Backer and Sramek 1979, 1987). These measurements make use of the interferometer phase, and have allowed us to determine the source centroid with the precision of a small fraction of the apparent source size both at 2.7 GHz (1979) and at 5.0 GHz (1987).

In the 2.7 GHz experiment at NRAO Green Bank we have compared the results to simultaneous measurements at 8.1 GHz and have found identical results at the level of 10 mas. This limit on the possible deviation of the scattered image centroid at one frequency from that at a higher (and less scattered) frequency is shown by the plus mark in Figure 1. This is a strong limit on refractive effects along this line of sight unless we are just unlucky in finding a refraction null between these two frequencies at the current epoch. Rickett (1986) suggests that the refractive time scale at 2.7 GHz is 20 years, or significantly greater than our observation interval.

In the 5.0-GHz experiment at the VLA we can place a limit of 1 mas on the epoch-to-epoch, *i.e.*, year-to-year, meandering of the source centroid. This limit is also marked in Figure 1. In this case the refractive time scale is estimated to be 5 years, so that our null result is barely significant. If there is refractive meandering of the source, then our measured proper motion will be contaminated by a systematic error. We have now observed a linear motion of 35 mas, or 0.7 of the apparent image size. This motion is consistent with that expected for a body at rest in the center of the galaxy. Since we do not know the nature of this source, we are not justified in stating that it must be at rest and that there is no evidence for a refractive component in the measurement. Measurements at a higher frequency will be proposed to improve these results.

The proper motion measurements described are differential. Angular scattering is less for the reference sources than for SgrA*, but there could be an additional contribution to refractive meandering of the differential position from the reference sources. The refractive limits for a single path are even more stringent than stated above.

INTENSITY VARIATIONS

The first evidence for variations of the intensity of SgrA* were published by Brown and Lo (1982). These authors showed that the flux density at 2.7 GHz displayed a linear rise over 1.5 years from 0.35 to 0.50 Jy, while the intensity at 8.1 GHz was secularly stable at 0.6 Jy. At both frequencies they gave evidence for 15 per cent variations on shorter time scales. Zhao (1988) has collected observations from various VLA observations at 20cm and 6cm between 1981 and 1988. These data indicate a decline in the 20cm intensity by 50 per cent in seven years with smaller variations on shorter time scales, and a 30 per cent decline with large variations at 6cm.

The variations require the source size to be less than 10 and 1 pc for the slow and rapid time scales, respectively. This light travel time argument does not require the angular size discussed above to be extrinsic to the source.

Perhaps the slow variations are refractive as discussed by Rickett and others in this meeting. Direct application of Rickett's equation 10 with τ_R estimated at 3 years and $\theta = 100$ mas gives an unreasonable velocity of 1450 km s^{-1}. One firmly concludes that only the slow variations seen at the highest frequencies are consistent with refraction effects. The location of the scattering region along the line of sight to SgrA* will not affect this conclusion (Rickett 1986).

The dominant source of intensity variability must reside in the emission mechanism for this enigmatic object. The proper motion and VLBI results suggest that SgrA* is a supermassive object, $M \geq 200$ $M_\odot$. Its radio emission is then presumed to be a synchrotron corona around this object. The existence of intrinsic variations on various time scales is then not surprising. The change in character of the variations from long to short wavelengths (Brown and Lo 1982; Zhao 1988) is also not a surprise in view of the lack of a detailed model for the origin and evolution of the energetic particles.

I thank my colleagues with whom I have observed SgrA* and R. Romani for discussions and a critical reading of the manuscript.

REFERENCES

D. C. Backer, *Ap. J. (Lett.)*, **222**, L9, (1978).
D. C. Backer, and R. A. Sramek, *Ap. J.*, **260**, 512, (1982).

D. C. Backer, and R. A. Sramek, *Proc. Symposium in Honor of C. H. Townes*, ed. D. C. Backer, [AIP : NY], p. 163, (1987).
R. L. Brown, and K. Y. Lo, *Ap. J.*, **253**, 108, (1982).
W. A. Coles, R. G. Frelich, B. J. Rickett, and J. L. Codona, *Ap. J.*, **315**, 666, (1987).
T. R. Clifton, Ph. D. Thesis, University of Manchester (1986).
R. D. Davies, D. Walsh, and R. S. Booth, *M. N. R. A. S.*, **177**, 319, (1976).
C. R. Gwinn, J. M. Moran, and M. J. Reid, This publication (1988).
K. Y. Lo, M. H. Cohen, A. C. S. Readhead, and D. C. Backer, *Ap. J.*, **249**, 504, (1981).
K. Y. Lo, D. C. Backer, R. D. Ekers, K. I. Kellermann, M. Reid, and J. M. Moran, *Nature*, **315**, 124, (1985).
J. Marr, personal communication (1988).
P. G. Mezger and T. Pauls, IAU Symposium No. 84, ed. W. B. Burton, [D. Reidel : Dordrecht] p. 357, (1978).
S. R. Spangler and J. M. Cordes, This publication (1988).
B. J. Rickett, *Ap. J.*, **307**, 564, (1986).
L. F. Rodríguez, J. Cantó, and J. M. Moran, *Ap. J.*, **255**, 103, (1982).
J. Zhao, personal communication (1988).

ANGULAR BROADENING MEASUREMENTS OF THE SOURCES 1849+005 AND 2013+370

Steven R. Spangler
University of Iowa, Iowa City, Iowa 52242
and
James M. Cordes
Cornell University, Ithaca, New York 14853

ABSTRACT

Angular broadening measurements, with different instruments and for different purposes, are reported for the radio sources 1849+005 and 2013+370. The former source lies in a part of the sky characterized by heavy scattering. Observations with the Very Large Array at frequencies of 1.46 and 0.33 GHz were made of this object and 45 others in its vicinity. We were particularly interested in investigating the possibility that the supernova remnant G33.6+0.1 might be responsible for the enhanced scattering. Preliminary inspection of the data reveals at least one other source which is approximately as heavily scattered as 1849+005. An intriguing preliminary result is a possible increase of scattering in the galactic plane with decreasing galactic longitude. Our observations indicate that between 20 and 30 degrees of longitude, the 1.46 GHz scattering size is of the order of 1 to 2 arcseconds. The second source, 2013+370, is much less heavily scattered than 1849+005, and VLBI is needed to resolve the scattering disk. The line of sight to this object transfixes the Cygnus OB1 association. Our observations of this source were primarily motivated to measure the spectral index of interstellar density irregularities. The observations are consistent with a value of $\alpha = 3.79 \pm 0.05$, which supports a quasi-Kolmogoroff spectrum in the interstellar medium.

INTRODUCTION AND MOTIVATION

We are presently engaged in measuring interstellar scattering of extragalactic radio sources which are either at low galactic latitudes, or otherwise have lines of sight of interest. In this paper we present results on two sources which illustrate the type of information such observations may furnish about the interstellar medium.

There are two reasons for making angular broadening measurements. The first is that such measurements arguably furnish the best opportunity for measuring the spectrum of the interstellar density fluctuations. The measured interferometric visibility is directly related to the density power spectrum, provided conditions for the measurement of an ensemble average are met. The second reason is that VLBI and VLA angular broadening experiments provide us with the opportunity of mapping the galactic distribution of turbulence, and thereby identifying the astronomical objects which are the agitators. Such knowledge would be a powerful clue for determining the physical mechanisms responsible for generating these fluctuations.

The first of these issues is addressed by our observations of the source 2013+370, whereas the observations of 1849+005 are relevant to the second of the goals of angular broadening measurements.

© 1988 American Institute of Physics

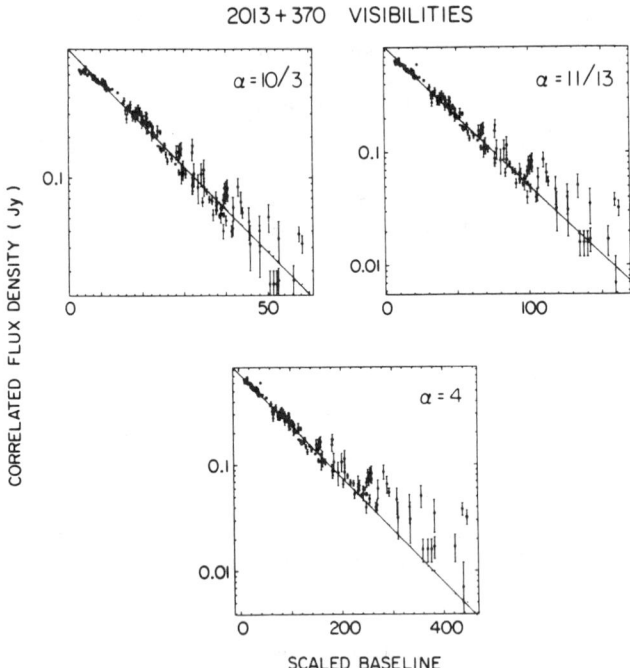

Figure 1: Radio source 2013+370. Correlated flux density versus baseline to the power $\alpha - 2$. Such a plot should be a straight line for the proper value of α. Plots are shown for α =3.33, 3.67, and 4. The Kolmogoroff value of 3.67 is superior to the other values. A least-squares fit yields a best value of 3.79.

VLBI OBSERVATIONS OF 2013+370

The angular broadening of 2013+370 was noticed during a search for enhanced scattering in the vicinities of supernova remnants[1]. It was noted that the broadening of 2013+370 was such as to be quite convenient for measurement with the United States VLBI network, and detailed VLBI observations at 1.66 GHz were subsequently made to furnish a measurement of the interstellar density spectrum.

The broadening image of 2013+370 is shown in Figure 2 of the paper by Spangler at this meeting. A least squares fit of the visibilities to a model power spectrum yields a spectral index, α of 3.79 ± 0.05, which indicates "quasi- Kolmogoroff" turbulence along the line of sight to this object. Figure 1 presents our evidence for this being a significant result. Show are correlated flux density versus baseline to the power $\alpha - 2$ for values of α =3.33, 3.67, and 4.

The density spectrum with α =3.33 provides a good fit to the long baseline data, but significantly overshoots the measurements on short baselines. The $\alpha = 4$ spectrum provides a good fit on the short baselines, but for long baselines the measured visibility significantly exceeds the model. This conclusion is enhanced by the realization that the most probable effect of correcting for finite source structure would be an upward revision of the measurements on long baselines, thus exacerbating the disagreement[2].

Finally, the Kolmogoroff value of $\alpha = 3.67$ is very close to the least squares fit value, and provides the best fit on all baselines.

From this result we conclude that the density turbulence along the line of sight to 2013+370 has a spectrum which is close or equal to the Kolmogoroff value. The refractive effects produced by such density turbulence would not be as pronounced as those observed in the dynamic spectra of pulsars.

VLA OBSERVATIONS OF 1849+005

This source was another of those observed for enhanced scattering by virtue of proximity to a supernova remnant[1]. This object presented an interesting dilemma. Although independent radio observations indicated that it is a compact extragalactic radio source, we were unable to detect fringes to the object, even on short VLBI baselines at 5 GHz. There were two possibilities for this nondetection; either the object was not a compact extragalactic radio source, or the source was so heavily scattered that it was completely resolved by the VLBI interferometers. For the latter, more interesting possibility, it would be necessary for the source to have a broadening size greater than 0.35 arcseconds at 1 GHz.

Sources of such angular sizes are in the domain of the VLA rather than VLBI, so we proposed observations with the former instrument at frequencies of 1.46 and 0.33 GHz. In addition, 45 other sources in the vicinity of 1849+005, chosen from a galactic plane survey by Garwood and Dickey[3], were observed.

There were two purposes for these observations. First, we wanted to verify the indication that 1849+005 is a highly scattered radio source. Second, we wanted to measure the scattering of sources adjacent to 1849+005 to see if the enhanced broadening of 1849+005 is attributable to the supernova remnant G33.6+0.1.

The observations were made last summer, and we are presently in the process of reduction, analysis and interpretation. The following remarks are therefore highly preliminary, but convey what seem to be some interesting results.

The extreme scattering of 1849+005 is confirmed. The measured angular size at 1.46 GHz is 0.5 arcseconds, consistent with the upper limit from the VLBI observations. The 333 MHz map is shown in Figure 2. The image is highly anisotropic, with a major axis of 13 arcseconds and a minor axis of 4. The harmonic mean angular size of 7 arcseconds is in approximate agreement with a λ^2 scaling of the size measured at 1.46 GHz. We conclude that we have detected interstellar scattering, with a scattering measure (path integral of the quantity C_N^2) 260 times that observed for 2013+370.

The next question to address is whether the source of this enhanced scattering is the nearby supernova remnant G33.6+0.1, another object, or more widely distributed turbulence. This matter can be addressed by measuring the angular diameters of nearby sources at both 333 and 1465 MHz, determining the the interstellar scattering contribution, and seeing whether 1849+005 is dominant in this region.

In practice it is difficult to thoroughly address this point. Only about seven of the sources we observed have adequate maps at 333 MHz, a result of the weakness of the sources and the relative insensitivity of the VLA for galactic plane observations. Our preliminary results give us the angular size of three sources within 5 degrees of 1849+005, and in the galactic plane. The largest scattering size is observed for 1849+005, and indeed for two of the objects the measured angular size does not scale with λ^2, so there

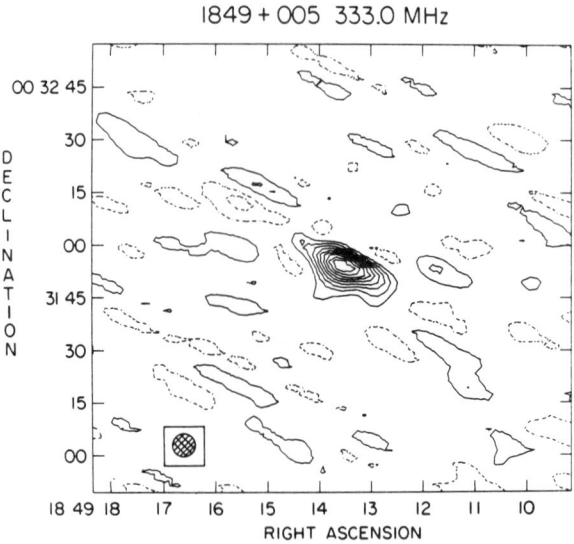

Figure 2: VLA map of the radio source 1849+005 at a frequency of 333 MHz. The source is clearly resolved and anisotropic. Contours are in levels of 10 percent of the peak intensity.

is no evidence for interstellar scattering. This would seem to indicate that the line of sight to 1849+005 might be noteworthy with regard to scattering strength.

Further features of our observations, however, indicate that the line of sight to 1849+005, while more highly scattered than some of its neighbors, is not unique in this part of the sky. The 333 MHz angular size of the radio source 1855+031 is roughly half that of 1849+005.

Even more compelling evidence for *general* heavy scattering in this part of the sky is shown in Figure 3. From the original source roster of 44 sources, we have culled 16 objects which appear to be relatively compact extragalactic sources, and therefore suitable for angular broadening measurements. All sources are within 0.5 degrees of the galactic plane. In Figure 3 we plot the 1.46 GHz angular diameter versus galactic longitude. It should be noted that our 1.46 GHz angular sizes are *upper limits* to the scattering size; for some of the sources in the vicinity of 1849+005 we know from the dual frequency measurements that the scattering disk size is smaller than the 1.46 GHz size. Nonetheless, what is revealed in Figure 3 is a tendency for the sources inside of 30 degrees longitude to have angular diameters in excess of those outside of 30 degrees. The obvious interpretation is that for sources inside 30 degrees, the observed size is dominated by interstellar scintillation, whereas for higher longitudes the measured sizes are a combination of intrinsic and scattered contributions. This result, though preliminary, indicates that the scattering in this part of the galaxy is of the order of 1 arcsecond at 1.5 GHz.

Such scattering is larger than that observed for 1849+005, and would indicate that supernova remnants are not the dominant sources of interstellar scattering.

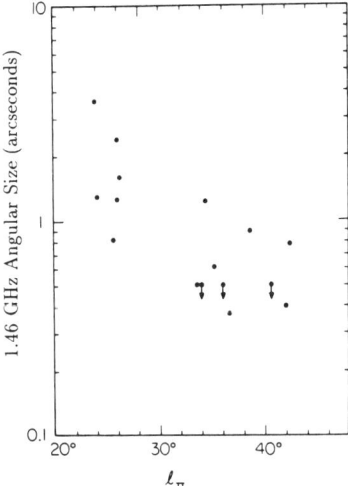

Figure 3: Angular size at 1.46 GHz versus galactic longitude for 16 sources which are probably compact extragalactic sources.

CONCLUSIONS

Angular broadening measurements have been made of two radio sources, addressing two aspects of interstellar scattering. VLBI observations at a frequency of 1.66 GHz of the radio source 2013+370 yield a measurement of α, the spectral exponent of the (assumed power law) interstellar density power spectrum of 3.79±0.05, supporting a quasi-Kolmogoroff spectrum along the line of sight to this source.

Observations with the VLA at 0.33 and 1.46 GHz of the source 1849+005 seem to have identified a region of particularly strong interstellar scattering close to the galactic plane, which shows pronounced changes on small angular scales. Further study of this region might help identify the mechanism for generation of the density turbulence.

ACKNOWLEDGEMENTS

This research was supported at the University of Iowa by grants NAGW-806 and 831 from the National Aeronautics and Space Administration, and at Cornell University by the Alfred P. Sloan Foundation, NSF grant AST-8520530, and the National Astronomy and Ionosphere Center. We sincerely thank Robert Garwood and John Dickey of the University of Minnesota for providing us with positions of sources from their survey prior to publication.

REFERENCES

1. S.R. Spangler, R.L. Mutel, J.M. Benson, and J.M. Cordes, Astrophys. J., 301,312,(1986).

2. S.R. Spangler and J.M. Cordes, Astrophys. J., in press, (1988).

3. R. Garwood and J. Dickey, Astron. J., submitted, (1988).

VLBI MEASUREMENTS OF INTERSTELLAR TURBULENCE SPECTRA

R. L. Mutel
University of Iowa, Iowa City, IA 52242

J. F. Lestrade
Bureau des Longitudes, Paris 75015, France

ABSTRACT

VLBI observations of the scattered images of two compact radio sources whose lines of sight pass through the Cygnus superbubble region were analyzed to infer the spatial spectrum of intervening turbulence. For the quasar 2005+403, the phase structure function varies quadratically with projected baseline for $b \lesssim 4 \cdot 10^8$ cm., but then flattens to a slope near zero. This is inconsistent with Type B ($\beta > 4$) power-law turbulence unless the outer scale is $L_0 \sim 2 \cdot 10^9$ cm. It is consistent with Type A ($\beta < 4$) spectra with an inner scale $l_0 \sim 2 \cdot 10^9$ cm., and a spectral index $2.0 \lesssim \beta \lesssim 2.5$. For the compact double source 2050+364, an upper limit of $\Delta\theta \lesssim 3$ mas was obtained at 0.61 GHz for the differential image wander of components, much less than the measured diffractive scattering disk $\theta_d = 25$ mas This is consistent with Type A spectra but not Type B, for which one expects $\Delta\theta \sim \theta_d$.

INTRODUCTION

Interstellar turbulence causes random phase and amplitude fluctuations of radio waves as they propagate through regions of varying refractive index. Analysis of these fluctuations at the earth enables one to deduce the average strength and spectrum of turbulence along the line of sight. Studies of the second moment of the scattered field (mutual coherence) are particularly useful, since they are mathematically more tractable than intensity measurements which involve fourth moment calculations.

The nature of the scattered image depends strongly on whether the scale size of turbulent cells is smaller or larger than the refractive or 'multi-path' scale $r_{mp} = z\theta_d$ where θ_d is the diffractive image size and z the propagation distance. For turbulence with irregularity scale sizes smaller than r_{mp}, the image consists of a smooth, nearly circular diffractive scattering disk. For larger scales, several refractive effects appear: image wander, multiple sub-images, large ellipticity, and intensity-image correlations. For power-law spectra, refractive effects appear when the spectral index $\beta > 4$; these are called 'steep', or Type B spectra. Spectra with $\beta < 4$ are called 'shallow' or Type A spectra. Although some evidence exists for refractive focussing based on periodic band structure in pulsar dynamic spectra[1], there is as yet *no* evidence for Type B spectra based on scattered images.

© 1988 American Institute of Physics

In this paper, we discuss observational tests of the spatial spectrum of turbulence assuming a simple power-law model with spectral index β and a finite inner and outer scale. The tests are applied to VLBI observations of two highly scattered compact extragalactic sources whose lines of sight lie within the Cygnus 'superbubble' region[2]. The region is known to contain large amounts of turbulent plasma, since several other nearby sources are also heavily scattered[3]. The location of the sources with respect to the superbubble is shown in Figure 1.

The compact double source 2050+364 provides a practical test of image wander since it consists of two nearly equal, highly scattered compact components separated by only 59 mas This small separation allows a very high accuracy differential measurement of the component separation at multiple epochs without the need for milliarcsecond absolute positions. The high redshift quasar 2005+403 was observed using a sensitive transcontinental MkIII VLB array over several days. The sensitivity was sufficient to measure the departure from a gaussian brightness distribution on long baselines, as expected from power-law turbulence spectra. By constructing the phase structure function from the measured visibilities, one can look for spectral breaks as a function of projected baseline length. For simple power-law turbulence spectra, the spatial frequency of the break is simply related to either the inner ($\beta < 4$) or outer ($\beta > 4$) scale.

VISIBILITIES AND STRUCTURE FUNCTIONS: 2005+403

The visibility Γ measured by a interferometer with baseline $\vec{b}$ can be written as:

$$\Gamma(\vec{b}) = < E(\vec{x}) \cdot E^*(\vec{x}+\vec{b}) > = e^{-\frac{1}{2}D_\phi(\vec{b},z)} \qquad (1)$$

where the brackets indicate an ensemble average and $D_\phi(\vec{b}, z)$ is the phase structure function. In practice, an ensemble average cannot be achieved in a single observation since the time required for the largest turbulent patches to traverse the line of sight is much longer than the observation.

Narayan and Goodman[5] have recently analyzed the differences between the measured and the ensemble averaged visibility as a function of integration time and turbulence spectrum. They consider power-law turbulence with a bulk velocity v across the line of sight. For coherent integration times normally used during VLBI observations (longer than the 'diffraction' time ($t_{diff} = z\lambda/vr_{mp} \sim$ few minutes) but shorter than the 'refraction' time ($t_{ref} = r_{mp}/v >$ weeks) the difference between the measured and ensemble averaged visibility is small for Type A spectra when $t_{coh} > r_{diff}/v = \lambda/2\pi v\theta_d$. For Type B spectra, the difference is quite large and nearly constant. The resulting image has a fractal structure with many sub-images. For both types of spectra, if the bandwidth of the observation exceeds the decorrelation bandwidth ($\Delta\nu_{dc} = c/z\pi\theta_d^2$), sub-image structure will be smoothed. (For Type B spectra, the diffractive angular size θ_d should probably be replaced by the sub-image size.)

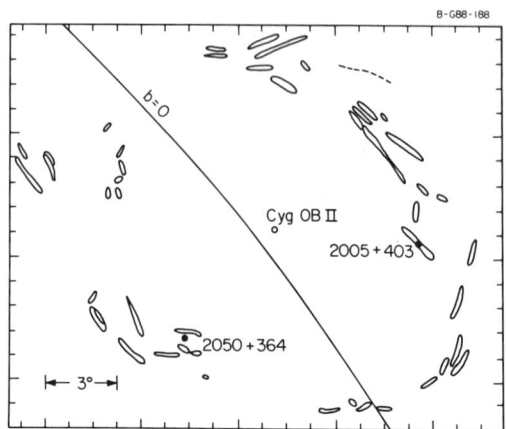

Figure 1 Location of the scattered sources 2050+364 and 2005+403 in the Cygnus superbubble region. The oval shaped structure is a region of elongated optical filaments surrounding the Cygnus OB2 association.[2,4]

For the observations discussed here, the coherent integration time was 4 min. and the bandwidths were 2 MHz (2050+364) and 26 MHz (2005+403). In both cases, $t_{coh} > t_{diff}$ (for $v > 10^6$ cm/sec). However, the observing bandwidth was greater than the decorrelation bandwidth ($\Delta\nu_{dc} \sim 10^{4-5}$ Hz) in both cases. This will quench sub-image structure. Nevertheless, in the following we will assume that the measured visibility data is a good estimate of the ensemble averaged visibility.

The phase structure function is related to the spatial spectrum of electron density turbulence $P_n(\vec{q})$ by:

$$D_\phi(\vec{b}, z) = 4\pi r_e^2 \lambda^2 \int_0^\infty P_n(\vec{q}, z, q_z = 0)(1 - \cos(\vec{q} \cdot \vec{s}) d^3q \qquad (2)$$

For isotropic, uniform turbulent region of thickness Δz, this reduces to:

$$D_\phi(\vec{b}) = 8\pi^2 r_e^2 \lambda^2 \Delta z \int_0^\infty P_n(q)[1 - J_0(qb)] q \, dq \qquad (3)$$

One often assumes a simple power-law model for the spatial spectrum of turbulence with an inner scale l_0 and an outer scale L_0:

$$P_{n_e}(q) = C_n^2 q^{-\beta} \qquad L_0^{-1} < q/2\pi < l_0^{-1} \qquad (4)$$

Cordes, et al.[6] have calculated structure functions for power-law turbulence using both Type A and B spectra. For Type A spectra, the phase structure function is piece-wise power-law with two characteristic break points:

$$D_\phi(b) \propto \begin{cases} b^2 & 2\pi b < l_0 \\ b^{\beta-2} & l_0 < 2\pi b < r_{mp} \\ b^0 & 2\pi b > r_{mp} \end{cases} \quad (5)$$

For earth-based VLBI observations, it is very unlikely that the multi-path break could be observed, since $r_{mp} \sim 10^{13-15}$ cm. at centimeter or meter wavelengths.

For Type B spectra, the structure function is also piece-wise power law, but with a break only at the outer scale:

$$D_\phi(b) \propto \begin{cases} b^2 & 2\pi b < L_0 \\ b^0 & 2\pi b > L_0 \end{cases} \quad (6)$$

Note that since the phase structure function can be constructed from the observed visibilities (eqn. 1), it may be possible to *directly* measure the inner scale for Type A turbulence by measuring the baseline at which the slope changes from 2 to $\beta - 2$.

The compact radio source 2005+403 is a high redshift (z =1.73) quasar with a flat spectrum at centimeter wavelengths. Don Backer (Heidelberg VLBI conference, 1978) first showed that it is one of the mostly highly scattered sources known, with diffractive scattering size of $\theta_d = 80$ mas at 1 GHz. We observed this source during November 1986 using a 6 element MkIII VLBI array at 5 GHz. We observed for 2 hours on each of three days with the same (u, v) coverage. The visibilities were very low on long baselines, but easily detectable with the sensitive MkIII system. The visibility amplitude and derived phase structure function as a function of projected baseline is shown in Figure 2.

The structure function is quadratic for baselines $b < 4 \cdot 10^8$ cm. but there is a clear break to a nearly flat slope for longer baselines. This behavior was seen on data from all three days, and was also found using the MkII data of Fey, *et al.*[3] taken six months earlier. Since it is very unlikely that the outer scale is as small as $\sim 10^9$ cm., the break appears to rule out Type B turbulence spectra. For Type A turbulence, a break is expected at $l_0/2\pi$ (eqn. 5), but the spectral index β must be nearly 2, since the observed slope beyond the break is nearly zero.

Examination of the closure phases on long triangles show a large, systematic deviations from 0°, indicating that intrinsic source structure is becoming visible on long baselines. (Multiple sub-images caused by scattering will be *coherent* and will therefore have zero closure phase and unity closure amplitude.) This resolution of the source structure will bias the phase structure function to *higher* values on long baselines, and so cannot explain the flattening.

We conclude that the observations are inconsistent with Type B turbulence, but could be explained by a Type A spectrum with an inner scale $l_0 \sim 2 \cdot 10^9$ cm. and a spectral index $\beta \lesssim 2.5$.

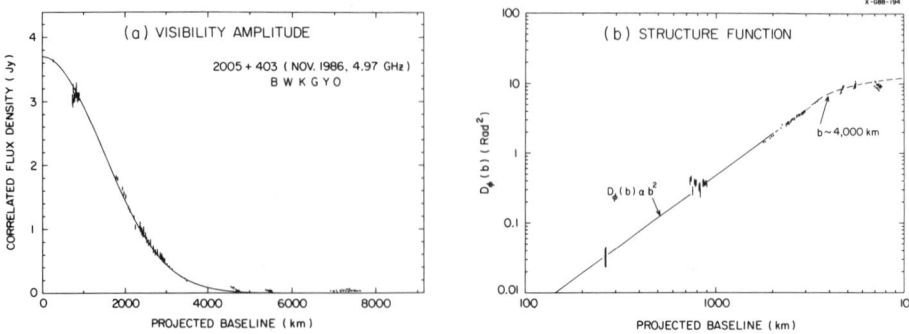

Figure 2 (a) Visibility amplitude and (b) Phase structure function for the source 2005+403 at 5 GHz. The structure function has a quadratic dependence for $b < 4 \cdot 10^8$ cm. but flattens rapidly for longer projected baselines. The visibility amplitude has been corrected for positive bias due to finite signal-to-noise ratio.

IMAGE WANDER

Very large turbulent eddies ($l > r_{mp}$) act as lenses which refract the scattered image so that it wanders from its average position. The mean wander angle $\Delta \theta_{ref}$ is a small fraction of the diffractive scattering disk for Type A spectra, but can be much larger for Type B spectra because of the preponderance of large irregularities. Romani, et al.[7] give numerical expressions for the expected wander angle and its time derivative in terms of the scattering strength, observing wavelength, and propagation distance for spectral indices $\beta = 3.7$, 4.0, and 4.3. Recasting these expressions in terms of the diffractive scattering angle, we obtain:

$$\Delta \theta_{ref}(mas) = \begin{cases} 0.11 \theta_d^{0.67} \lambda_{cm}^{0.13} \Delta z_{kpc}^{0.63} & \beta = 3.7 \\ 0.28 \theta_d (\mu/\gamma) & \beta = 4.0 \\ 0.77 \theta_d^{0.85} \Delta z_{kpc}^{0.15} s^{0.15} & \beta = 4.3 \end{cases} \quad (7)$$

$$\Delta \dot{\theta}_{ref}(\mu as/day) = \begin{cases} 0.4 \theta_d^{-0.33} \lambda_{cm}^{0.73} \Delta z_{kpc}^{-1.20} v_7 & \beta = 3.7 \\ 22 \gamma^{-0.5} \Delta z_{kpc}^{-1.0} v_7 & \beta = 4.0 \\ 36 \Delta z_{kpc}^{-1.0} v_7 & \beta = 4.3 \end{cases} \quad (8)$$

where γ and μ are dimensionless logarithmic factors close to unity, v_7 is the relative velocity transverse to the line of sight in units of 10^7 cm/sec, and $s = v_7 t_{yr}$ is the spatial lag.

In order to test for image wander, one needs to measure absolute source positions between observing epochs with an error less than the diffractive scattering disk. In practice, the required accuracy ($\lesssim 10$ mas at 1 GHz) is not possible with present VLBI arrays because of phase fluctuations caused by instrumental and ionospheric instabilities. An alternate scheme which avoids these problems is to measure the separation of a widely separated double source at several observing epochs. Since this is a *differential* measurement, determination of absolute position with milliarcsecond accuracy is unnecessary. Differential wander will result from turbulent scale sizes in the range $r_{mp} \lesssim l \lesssim z\Theta$ where Θ is the component separation and z the distance to the screen. Smaller irregularities will cause diffractive scattering, while much larger sizes will cause both components to wander in the same manner.

The compact double source 2050+364 is a excellent choice for such a measurement. It consists of two nearly equal compact components separated by 59 mas. It has been mapped at 0.6, 1.6, and 5.0 GHz using VLBI arrays[8,9]. Mutel and Hodges[9] showed that component sizes scale approximately as λ^2, which they suggest is caused by interstellar turbulence from a scattering screen ~ 0.5 kpc thick associated with the Cygnus superbubble. The diffractive image size is $\theta_d = 25$ mas at 0.61 GHz, smaller than the component separation, but much larger than the differential position accuracy of the interferometer (~ 3 mas).

Since a VLBI map at 0.61 GHz is available at only one epoch, we have used the component separation determined from the 5.0 GHz map[8] to determine the intrinsic component separation, $\Theta_0 = 59 \pm 2$ mas (the eastern component consists of three sub-components spread over ~ 5 mas, so a mean position was used). To determine the component separation and uncertainty at 0.61 GHz, we examined the reduced χ^2 for varying separations of a two component gaussian model fit to the visibility data. The result was $\Theta = 59 \pm 3$ mas, where the errors refer to 1σ variation. This implies a wander of:

$$\Delta\theta_{ref}(mas) = \Theta - \Theta_0 = 1 \pm 4 \tag{9}$$

Using equation (7), the expected mean wander angle is:

$$\Delta\theta_{ref}(mas) = \begin{cases} 1.0 & \beta = 3.7 \\ 7.0 & \beta = 4.0 \\ \sim 10.7 & \beta = 4.3 \end{cases} \tag{10}$$

where we have assumed $\gamma = \mu = s = 1$ and taken $\Delta z = 0.5$ kpc. The measured wander clearly favors Type A turbulence. However, there are at least two reasons why the observed wander may be less than expected: (*a*) the expressions given in equation (7) refer to wander as measured in an absolute frame; our differential measurement suppresses contributions from scales larger than $z\Theta_0$, and (*b*) it is possible that any single measurement of the component separation could lie close to the mean (intrinsic) value by coincidence. The latter possibility can easily be eliminated by multiple epoch observations at 0.6 GHz.

We conclude that both observations favor Type A models of interstellar turbulence.

REFERENCES

1. Hewish, A., Wolszcan, A. and Graham, D.A., *M. N. R. A. S.*, **213**, 167, (1985)
2. Bochkarev, N.G. and Sitnik, T.G. *Astr. Sp. Sci.*, **108**, 237, (1985).
3. Fey, A.L., Spangler, S.R., and Mutel, R.L., *This conference.*
4. Dickel, H.R. , Wendker, H. and Bieritz, J.H. *Astr. Ap.*, **1**, 270, (1969)
5. Narayan, R. and Goodman, J., *submitted to M.N.R.A.S.* (1988)
6. Cordes, J.M., Pidwerbetsky, A., and Lovelace, R.V.E., *Ap. J.*, **310**, 737, (1986).
7. Romani, R.W., Narayan, R., and Blandford, R., *M. N. R. A. S.*, **220**, 19, (1986)
8. Mutel, R.L., Hodges, M.W., and Phillips, R.B., *Ap. J.*, **290**, 86, (1985)
9. Mutel, R.L. and Hodges, M.W., *Ap. J.*, **307**, 472, (1986)

Interstellar Scattering of Radiation from H_2O Masers in W49 and Sgr B2

C.R. Gwinn, J.M. Moran and M.J. Reid (Center for Astrophysics)

Abstract

Strong limits on the presence of large-scale ($\approx 10^{13}$ cm) density fluctuations in the interstellar plasma are obtained from the wander of H_2O masers in Sgr B2 and W49 about constant-velocity trajectories, when compared with the sizes of the masers' scattering disks. These limits, and the scaling of correlated flux density with baseline length for maser features in W49, are consistent with a spatial power-law spectrum of density fluctuations in the interstellar plasma with an index of about the Kolmogorov value, 3.67. Refractive scattering for these two lines of sight is small.

Introduction

At high spatial and spectral resolution, H_2O masers are observed as clusters of tens or hundreds of intense, compact spots distributed over a region of a few arc sec. For W49, Sgr B2, and many other masers more distant than 1 kpc, the minimum size of maser spot within a cluster is set by interstellar scattering. The maser spots within the cluster show relative proper motions as well as a range of Doppler shifts. Both proper motions and Doppler shifts reflect physical motions. The variation of the motions about constant-velocity trajectories limits the strength of large-scale density fluctuations in the interstellar plasma, which can produce such variations by refraction.

Observations and Data Analysis

We made a series of three epochs of VLBI observations of H_2O masers in W49 and Sgr B2 over a period of seven months from November 1981 to January 1982. For the Sgr B2 observations we used a network of four radiotelescopes: Haystack, Green Bank, a single VLA antenna, and Owens Valley. For the observations of W49, the Onsala antenna was included as well. All stations recorded left-circular polarization (IEEE convention) with the Mk II VLBI system at the H_2O maser frequency of 22.235080 GHz. For Sgr B2, we recorded a 2 MHz bandpass, time multiplexed to cover 6 MHz, shifted to an LSR Doppler velocity of 56 km sec^{-1}. For W49, the data considered here were from a 2 MHz bandpass centered at a velocity of -5 km sec^{-1}, only a small fraction of the bandwidth of maser emission.

The data were amplitude and phase calibrated as described by Reid et al. (1980). Telescope gains were determined from spectra derived from the autocorrelation functions and were used to calibrate the cross-power spectra. The cross-power spectra were corrected for known propagation and earth orientation differences between epochs, which introduce phase changes. The phases of a bright, isolated feature in each cluster were subtracted from the phases of the cross-power spectra, removing residual clock and atmospheric variations. We estimate that errors in the positions of masers relative to the reference features, due to receiver noise and errors in the models, are less than 10 microarcseconds (μas).

We used the fringe-rate mapping method (see Thompson, Moran and

Swenson 1986, p. 384) to locate maser features to an accuracy of a few milliarcseconds (mas). We then fit a model directly to the (u,v) data to obtain the location, flux density and size of each maser feature. Details of the fitting are discussed by Gwinn et al. (1988). For each maser detected with sufficiently high flux density at all three epochs, we obtained a time series of positions on the sky, relative to the maser feature used as a phase reference. The time series showed motions which were very nearly constant-velocity. For both Sgr B2 and W49, these motions showed the cluster of masers to be expanding at tens of km sec^{-1}. We use the observed deviations from constant velocity to set limits on refraction by large-scale plasma fluctuations.

Source Sizes and Scattering

A variety of evidence indicates that the minimum sizes of the H_2O masers in Sgr B2 and W49 are due to interstellar scattering. Both sources show a sharp cutoff in maser size, at 0.3 mas (FWHM) for Sgr B2 and at 0.2 mas (FWHM) for W49 (Gwinn et al. 1988). Some closer masers have smaller minimum sizes; the tendency of size to increase with distance is suggestive of a propagation effect. OH masers in W49 and Sgr B2 have angular sizes about 300 times that

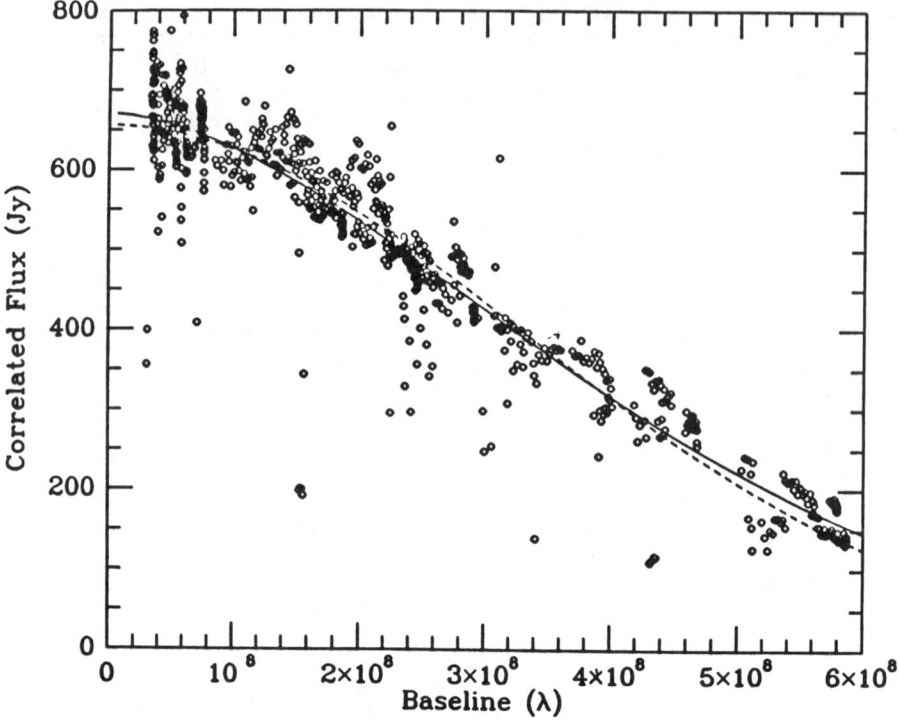

Fig. 1. Correlated flux as a function of baseline length for a bright H_2O maser feature in W49 with a Doppler velocity of -15.7 km sec^{-1}. Data points were phase and amplitude calibrated and time integrated for 120 sec. The best-fit scattering disk shape is indicated by the solid line; the best-fit Gaussian distribution by the dotted line.

of the H_2O masers in these regions, as expected for a scaling of angular size θ with wavelength λ as $\theta \propto \lambda^2$, as expected for interstellar scattering (Gwinn et al. 1988).

For H_2O masers in W49, the correlated flux density declines with baseline length as predicted for a power-law spectrum of density variations in the interstellar plasma, with an index near the Kolmogorov value of 3.67. Figure 1 shows the correlated flux density as a fuction of baseline length for a bright maser feature in W49 with an LSR velocity of -15.7 km sec^{-1}. The data have been edited to remove points for which the phase referencing or amplitude calibration failed. Scatter of the amplitudes due to their finite signal-to-noise ratio is less than 1 Jy; the apparent scatter reflects the effects of confusing features or calibration errors. The closure phases are consistent with zero. We also examined a maser feature with an LSR velocity of -0.8 km sec^{-1}, which shows closure phases of up to 25° on transatlantic baselines, indicating structure or large effects of confusing features. We fit the correlated flux density S_{12} of these bright features with a more complex model than that used for the (u, v) fitting. This model assumed that S_{12} took the form

$$S_{12} = S_0 \exp\left\{-\frac{1}{2}\left[\frac{\pi}{\sqrt{2\ln 2}}\theta_H B_\lambda\right]^{\alpha-2}\right\}$$

where the model parameters are S_0, the zero-spacing flux density of the maser feature; θ_H, a measure of the size of the feature, which is the FWHM of the scattering disk for $\alpha = 4$; and α, the spectral index of density fluctuations at scales near the baseline length (Salpeter 1967). The independent variable B_λ is the baseline in wavelengths. The (u, v) coverage for W49 was very elongated and nearly linear, so we are unable to estimate source elongation reliably. We measure the properties of the scattering disk along the axis of the (u, v) coverage, about 80° east of north. Table I summarizes the results of these fits. Quoted errors are estimated from the residual rms for the fits. The -0.8 km sec^{-1} feature has a larger size than the -15.7 km sec^{-1} feature, indicating possible structure or confusion from a nearby feature, in agreement with the closure phases. Nevertheless, the two values of α are consistent. As our best estimate for α, we take the fitted value for the -15.7 km sec^{-1} feature of $\alpha = 3.74 \pm 0.05$. This is slightly greater than the Kolmogorov value of $\alpha = 3.67$, but well below the Gaussian value of $\alpha = 4.00$.

Table I: Fitted Parameters for Two Maser Features

Parameter	Units	Symbol	-15.7 km sec^{-1} Feature	-0.8 km sec^{-1} Feature
Peak Flux	Jy	S_0	670.8 ± 4.7	845.6 ± 4.6
Size	mas	θ_H	0.244 ± 0.003	0.295 ± 0.002
Spectral Index		α	3.74 ± 0.05	3.69 ± 0.04

We were able to observe Sgr B2 only on baselines within the U.S., since Sgr B2 is not observable in Europe and the U.S. simultaneously. Even with the larger minimum size of the maser features in Sgr B2, the range of baseline lengths was too narrow to find α and θ_H simultaneously.

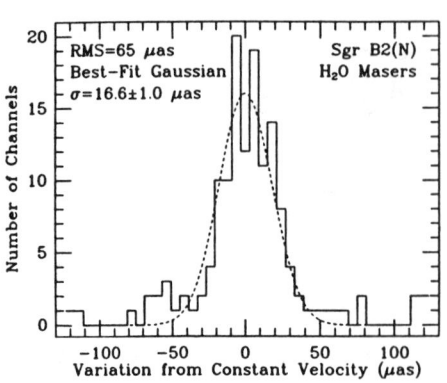
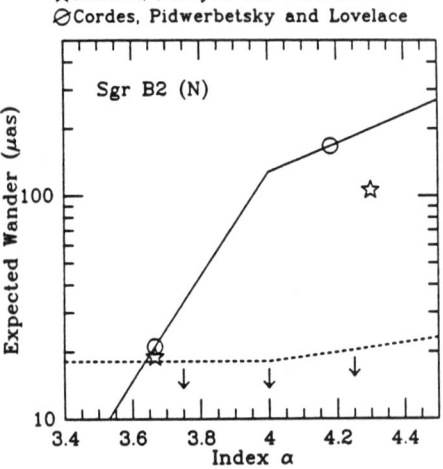

Fig. 2. Residual wander, after correction for the reduction in rms due to fitting constant-velocity motions, for masers in Sgr B2. The dashed line indicates the best-fit Gaussian distribution.

Fig. 3. Comparison of predicted wander as a function of α (solid line and points) with our observed upper limit (dashed line). The observational limit increases slightly for $\alpha > 4$ due to correlations of position with time, as discussed by Gwinn et al.(1988).

Maser Wander

Large-scale density fluctuations in the interstellar plasma can act as prisms, deflecting the radiation from a source so that it does not appear at its true position. To produce such deflections, the density fluctuations must subtend an angle of the order of the angular size of the source or greater, as seen from the earth. For both W49 and Sgr B2, this scale is about 10^{13} cm. Motion of the line of sight to the source relative to the scattering material, due to galactic rotation and the internal motions of the masers, produces angular deflection changes and apparent wander about the true trajectory. For power-law and other density spectra for which large-scale fluctuations appear all along the line of sight, one expects a Gaussian distribution for wander, since the different fluctuations combine incoherently to produce the net wander. We observe that masers move very nearly in straight lines, and use their observed deviations from straight-line motion to set an upper limit on the strength of large-scale density fluctuations in the interstellar plasma.

We used the 3-epoch series of positions described above to estimate maser wander. We fit linear motions in right ascension and declination to the time series and corrected the residuals for the reduction in rms due to the fit. Details of the analysis are described by Gwinn et al. (1988). Figure 2 shows a histogram of the residuals for Sgr B2. The residuals show a sharp peak, with a width of 16.6 ± 1.0 μas, as determined from the best-fitting Gaussian distribution. This result sets an upper limit of 18 μas (95% confidence limit) on any Gaussian process which contributes to wander. The distribution has broader wings than a Gaussian, with residuals as great as 600 μas. These contribute to the rms of 65 μas.

A prelimiary analysis of maser wander in W49 yields results similar to those obtained for Sgr B2. The best-fit Gaussian width is less than 45 µas, and the rms is less than 170 µas. Further analysis may reduce these preliminary limits.

Specific calculations by Cordes, Pidwerbetsky and Lovelace (1986) and Romani, Narayan and Blandford (1986) predict the scattering disk size and the expected wander from the properties of the medium for power-law density spectra. Using the distance to the source, the observing wavelength, and the velocity of the line of sight through the medium, we can predict the expected wander from the scattering disk size with their formulas. Figure 3 compares our observational upper limits on Gaussian wander with the theoretical predictions for Sgr B2. The observed wander is in good agreement with the predictions for the Kolomogorov spectrum for both theories. For W49, our preliminary limit on wander limits α to less than 3.85. The observed limits on wander are also consistent with a power-law spectra with long-wavelength cutoffs at less than 10^{13} cm, with α unconstrained.

Conclusions

The correlated flux as a function of baseline length for maser spots probes the index of the electron density spectrum at scales on the order of the baseline length, $\approx 10^8$ cm, while comparison of the scattering disk size and maser wander compares the strength of the density fluctuation spectrum at $\approx 10^8$ cm and $\approx 10^{13}$ cm. The scaling of correlated flux density with baseline length for W49 indicates that the index of the density fluctuation spectrum is about 3.7 at length scales near 10^8 cm. It also implies that there are significant density fluctuations over length scales of that order. The strength of the perturbations near 10^{13} cm, as probed by maser wander, is of particular interest, since strong fluctuations near this scale can produce intensity variations (Narayan and Blandford 1985). Our limits on maser wander show that, at least for these lines of sight, the large-scale fluctuations are too weak to produce dramatic intensity variations. The relative strengths of density fluctuations at 10^8 cm and 10^{13} cm are consistent with a Kolmogorov spectrum.

REFERENCES

1. J. M. Cordes, A. Pidwerbetsky and R. V. E. Lovelace, *Ap. J.*, **310**, 737 (1986).
2. C. R. Gwinn, J. M. Moran, M. J. Reid and M. H. Schneps, *Ap. J.* in press (1988).
3. M. J. Reid, A. D. Haschick, B. F. Burke, J. M. Moran, K. J. Johnston and G. W. Swenson, *Ap. J.*, **239**, 89 (1980).
4. R. W. Romani, R. Narayan and R. Blandford, 1986, *M.N.R.A.S.*, **220**, 19.
5. E.J. Salpeter, *Ap. J.*, **147**, 433 (1967).
6. A. R. Thompson, J. M. Moran and G. W. Swenson Jr., *Interferometry and Synthesis in Radio Interferometry* (New York: Wiley). (1986)

INTERSTELLAR ELECTRON DENSITY AND MAGNETIC FIELD IRREGULARITIES ON 0.001 TO 100 PARSEC SCALES

John H. Simonetti
Physics Department
Virginia Polytechnic Institute and State University,
Blacksburg, Va. 24061

James M. Cordes
Department of Astronomy, National Astronomy and Ionosphere Center
Cornell University, Ithaca, N.Y. 14853

ABSTRACT

Observations of the differences in Faraday rotation measures (RMs) of radio sources seen along different lines of sight yield information on interstellar electron density and magnetic field irregularities. RM observations of extragalactic sources imply the outer scale of the irregularities of either the electron density or the magnetic field, or both, is at least as large as 1 pc. Assuming the outer scale of the magnetic field is much smaller than 1 pc, the observed structure function of RM variations is consistent with an extrapolation, to large length scales, of the Kolmogorov spectrum of electron density irregularities implied by scintillation studies. If, on the other hand, the electron density outer scale is much less than 1 pc, we conclude that the fractional variations of electron density are much less than those of the magnetic field.

INTRODUCTION

Observations of the interstellar scintillation (ISS) of radio sources sample the interstellar electron density turbulence on length scales in the range $\sim 10^9$ to 10^{13} cm. Within this range of length scales these studies indicate the random irregularities have a power-law power spectrum, with a fairly well determined logarithmic-slope.[1] Just how far this form of the power spectrum extends is not well known. ISS studies may have already probed the smallest length scale present in the irregularities (the "inner scale"). The presence of drifting patterns in the dynamic spectra of the millisecond pulsar 1937+21 imply a lower limit on the largest length scale present (the "outer scale") of about 10^{14} cm.[2] Some theoretical models of the plasma turbulence responsible for the observed ISS and scattering indicate the outer scale may be 1-10 pc.[3] Finally, there exists tantalizing observational evidence for a power-law power spectrum of density variations in the interstellar medium (approximately matching the ISS spectrum) extending, without breaks, over the range $\sim 10^9$ cm to 10 pc, or more.[4] Nevertheless, length scales from 10^{14} cm to 10^{18} cm have been poorly studied.

Clearly, to see how well the ISS results extrapolate to different scales and to better constrain the outer scale, we need a way to sample, in detail, the turbulence on larger length scales than is probed by ISS. In this article we suggest one way to study the larger scales is to compare the rotation measures of linearly polarized extragalactic sources seen along slightly different lines of sight. We present an anayslsis of the data we have collected to date.

HOW ROTATION MEASURES ARE USEFUL

The utility of rotation measures is apparent when one considers the structure function of RM as a function of angular displacements on the sky. The observed position angle of radiation from a linearly polarized source is

$$\psi = \psi_0 + RM\ \lambda^2 \qquad (1)$$

where ψ_0 is the initial angle (at the source), λ is the observing wavelength, and the rotation measure RM is the integral

$$RM = \frac{e^3}{2\pi m^2 c^4} \int_0^L n\vec{B} \cdot d\vec{z} = \frac{e^3}{2\pi m^2 c^4} \int_0^L nB_z\ dz \qquad (2)$$

taken from the source to the observer (distance L), where n is the electron density and $\vec{B}$ is the magnetic field. For two sources separated by an angle $\delta\theta$, the RM structure function is defined as

$$D_{RM}(\delta\theta) = <[RM(\theta) - RM(\theta+\delta\theta)]^2> \qquad (3)$$

(angular brackets denote an ensemble average). For small $\delta\theta$, we take the path lengths through the interstellar medium of our Galaxy to be the same for both sources. In general, $D_{RM}(\delta\theta)$ will be the sum of two terms: a "statistical" term, due to the random irregularities in the medium, and a "geometric" term due to the relative orientation of the lines of sight through the mean field. As discussed previously[5] (Paper I), for $nB_z = <nB_z> + \delta(nB_z)$, where the power spectrum of the irregularities for wavenumber q is

$$P_{\delta(nB_z)} \propto q^{-\beta} \quad \text{for } (2\pi/s_{outer}) < q < (2\pi/s_{inner}), \quad 2<\beta<4, \qquad (4)$$

the statistical term in D_{RM} is $\propto \delta\theta^{\beta-2}$, for $\delta\theta$ between the inner and outer angular scales, and constant for $\delta\theta$ greater than the outer angular scale. For $4<\beta<6$, the statistical term is $\propto \delta\theta^2$ in the middle range of scales. Observations of a set of linearly polarized sources through a limited region of the interstellar medium, and with various separations on the sky, can be used to estimate the ensemble RM structure function. If the geometric term of D_{RM} is negligible, we may be able to discern β and the outer scale.

THE OBSERVED ROTATION MEASURE STRUCTURE FUNCTION

Extragalactic radio sources are in some respects ideal for this observational program. Numerous, and found in all directions, in many cases they can be resolved into separate components, thus enabling inter- and intra-source RM differencing. One drawback is their intrinsic contribution to RM, but this contribution is usually small. For line-of-sight distances L = 1 kpc through the interstellar medium (ISM) of our Galaxy, angular separations between individual sources of $\delta\theta$~1-10° correspond to transverse length scales s~20-200 pc; 10"-10' separations between components correspond to s~0.05-3 pc.

In Paper I we used published RM values to estimate RM structure functions for $\delta\theta$>1° in the direction of the North Galactic Pole (NGP; b > 60°), and toward a region of large, negative rotation measures (Region 1; 70°<l<110°, -45°<b<-5°). In a subsequent paper[6] we reported VLA observations of the RMs of individual source components in these directions, and through the Galactic Plane adjacent to Region 1 (GP; l~90°, |b|<10°). Figure 1a displays these data. The points connected by lines are averages computed for $\delta\theta$ bins of a few degrees using the RM values from the literature. The isolated points are RM differences between components of individual sources.

RANDOM INTERSTELLAR IRREGULARITIES DOMINATE THE RESULTS

The observed RMs have contributions from all media along the line of sight containing thermal electrons and magnetic field. Careful study of each source is necessary to select those for which only the ISM of our Galaxy is dominant. Reasonable strategies include checking the source depolarization with increasing λ to exclude sources where thermal electrons are mixed within the emission region, and using only components which are well separated and thus lie outside any ISM of the parent galaxy. Perhaps the safest strategy is to regard the NGP results as a "baseline" since RMs are generally quite low in that direction ($<\vec{B}>$ is perpendicular to the line of sight?). Assuming the sources and any contribution to RM irregularities from beyond our ISM are statistically similar for all directions, RM structure function values substantially greater than those for NGP sources represent contributions from our ISM.

This interpretation of the data implies that the results for Region 1, at least on the degree scales, are determined by our ISM. The same statement can apparently be made for the GP region, although more data are necessary.

Assuming reasonable parameters, the geometric term in D_{RM} is negligible for all directions. Solid and dashed lines in Figure 1a represent the geometric term when $<\vec{B}>$ is perpendicular and parallel to the line of sight for $<n> \approx 0.025$ cm^{-3} and $<B> \approx 3$ μG. L is likely to be about 0.5 kpc for lines of sight toward the NGP.

The observed RM structure function for Region 1 and GP is apparently dominated by the random irregularities in our ISM.

ELECTRON DENSITY AND MAGNETIC FIELD OUTER SCALES: TWO EXTREME CASES

For both Region 1 and GP it is likely that L = 1 kpc at least. Since there are no signs of an outer scale in either case, it

Diffraction Phenomena

appears the outer scale of electron density or magnetic field irregularities (s_n or s_B), or both, must be at least 1 pc. One possible objection to this conclusion is that RM variations sampled at these angular scales may be due to "deterministic" structures within the ISM (HII regions, shock fronts, etc.) that are not part of the "turbulent" power spectrum responsible for ISS. This point needs to be explored further, probably by trying to fill in the gap between the RM-probed scales and the much smaller ISS scales.

While it might be true that electron density and magnetic field variations will go hand in hand at any scale, and therefore both outer scales may be large, it is useful to explore the two extremes $s_n \ll s_B \sim 1$ pc, and $s_B \ll s_n \sim 1$ pc. Consider the latter case first.

If $s_B \ll 1$ pc, the observed RM structure function for Region 1 must be due to electron density irregularities on parsec scales. From ISS work, the power spectrum for electron density variations in the "diffuse" turbulent medium sampled by lines of sight $|b|>10°$ is

$$P_{\delta n} = C_n^2 \, q^{-\alpha}, \qquad \alpha \approx 4. \tag{5}$$

Upon extrapolation of this spectrum to parsec scales, we find

$$D_{RM}(\delta\theta) = 91 \, \frac{C_n^2}{10^{-3.5} \, m^{-20/3}} \, \frac{<B_z>^2}{(\mu G)^2} \left[\frac{L}{kpc}\right]^{8/3} \left[\frac{\delta\theta}{degrees}\right]^{5/3} \, rad^2 m^{-4} \tag{6}$$

using $\alpha = 11/3$ (the Kolmogorov value). Equation (6) is plotted as a solid line in Figure 1b. Also shown is the resulting line for $\alpha = 4$. It is clear that the $\alpha = 11/3$ result, an extrapolation from length scales some six to ten orders of magnitude smaller than those plotted, is quite consistent with the data for Region 1. The individual component data may be displaced from the line due to measurement error bias or contributions to RM differences beyond our ISM. The observed variations for GP are stronger than for Region 1, as they should be, to be consistent with ISS results for $|b|<10°$.

Consider the case $s_n \ll 1$ pc. Since $s_n = L\delta\theta_n$ for $s_n \ll L$, and

$$<\delta n^2> = \int_{2\pi/s_n}^{\infty} C_n^2 \, q^{-\alpha} \, 4\pi q^2 dq, \tag{7}$$

we find ($\alpha = 11/3$),

$$\delta\theta_n = 0.4° \left[\frac{C_n^2}{10^{-3.5} \, m^{-20/3}}\right]^{-3/2} \left[\frac{<n>}{0.025 \, cm^{-3}}\right]^{3} \left[\frac{L}{kpc}\right]^{-1} \left[\frac{<\delta n^2>}{<n>^2}\right]^{3/2} \tag{8}$$

Taking $s_n \ll 1$ pc ($\delta\theta \ll 0.4°$) must therefore imply $<\delta n^2>/<n>^2 \ll 1$. On the other hand, for Region 1 the observed variance of RM over the set of sources is > the square of the mean RM. The mean $\vec{B}$ field in that direction is nearly along the line of sight (therefore the large values of RM). These two facts imply $<\delta B_z^2> \approx <B_z>^2$, and $<\delta B^2> \approx <B>^2$. Therefore, $s_n \ll 1$pc apparently leads to the result $<\delta n^2>/<n>^2 \ll <\delta B^2>/<B>^2$. A possible objection to this conclusion lies in our assumption of $<n> = 0.025$ cm^{-3}, a value derived from pulsar dispersion measures. It may not be true that local values of the mean electron density are the same as the line-of-sight average obtained from the dispersion measure.

CONCLUSIONS

The RM data for extragalactic sources imply the turbulent power spectrum derived from ISS studies may extend to an outer scale of at least 1 pc. Whether this conclusion holds up will be seen upon the collection of more RM data. A particularly important range of scales can be probed by comparing the rotation measures of the individual components of sources.

These observations can also be used to study turbulence in different regions of the ISM. Our work indicates turbulence is stronger in the plane, in agreement with ISS observations. We are currently testing the idea that turbulence may be stronger near supernovae, which may ultimately be responsible for the irregularities that cause ISS.

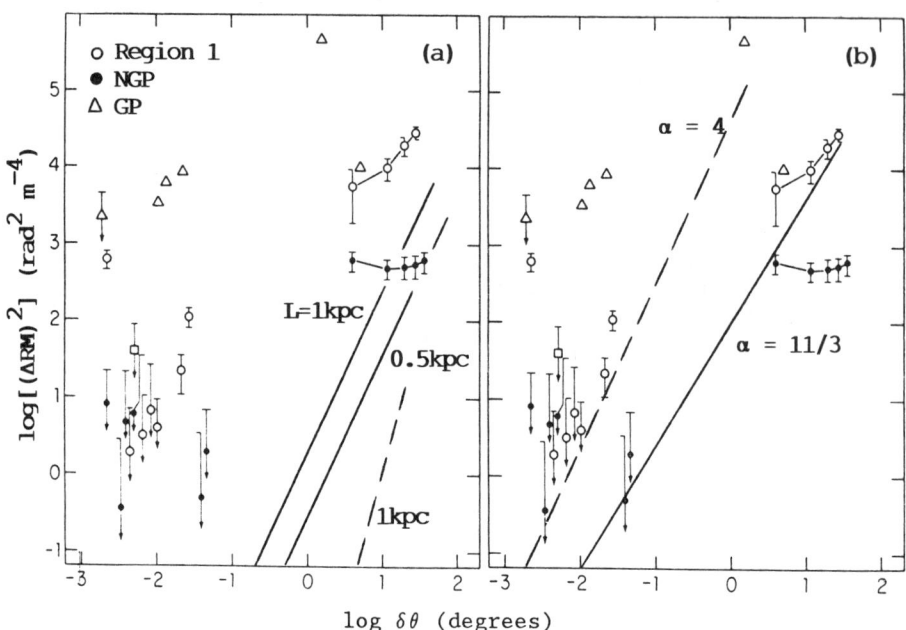

Fig. 1. Observed rotation measure structure function.

1. J.M. Cordes, J.M. Weisberg, V. Boriakoff, Ap.J. 288, 221 (1985).
2. J.M. Cordes, A. Pidwerbetsky, R.V.E. Lovelace, Ap.J. 310, 737 (1986).
3. I. McIvor, M.N.R.A.S. 178, 85 (1977).
4. J.W. Armstrong, J.M. Cordes, B.J. Rickett, Nature 291, 561 (1981).
5. J.H. Simonetti, J.M. Cordes, S.R. Spangler, Ap.J. 284, 126 (1984).
6. J.H. Simonetti, J.M. Cordes, Ap.J. 310, 160 (1986).

V. REFRACTION PHENOMENA AND LARGE SCALE STRUCTURE

Refractive Effects in Pulsar Dynamic Spectra

Y. Gupta and B. Rickett, University of California, San Diego
and
A. Lyne, Jodrell Bank

Abstract: Pulsar dynamic spectra exhibit several features which can be attributed to refractive interstellar scintillation (RISS). These include a characteristic drift of the features which shows changes on time scales of weeks to months and the much rarer occurrence of "fringe like" patterns. The drift of the features can be explained by a simple frequency dependent refractive shift of the diffractive intensity pattern. The two dimensional correlation of the intensity of the diffractive pattern can be modified to include the refractive effect. In this paper we fit the correlation function of the dynamic spectra with different models and study the evolution with time of the parameters of the model for several pulsars. The current results do not show consistent evidence for the simple refractive picture.

Pulsar dynamic spectra[3] are records of the received pulsar intensity as a function of time and frequency. Fig. 1 shows a typical example. As can be seen, the features have a characteristic drift in time versus frequency. This drift changes slowly with time. Typical time scales for significant changes in the slope are found to be of the order of several weeks to months. Fig. 2 shows autocorrelations of the spectra from two observations of P0823+26 one hundred days apart. Such changes of the drift are seen quite often for some of the pulsars in our data set, which typically has 12 observations of a pulsar over a 14 month period.

On some occasions, the dynamic spectra show intensity features grouped into regularly spaced bands, very much like a fringe pattern produced by interference.[5,6] Fig. 3 shows an example from the pulsar 2016+28. However, the occurrence of such patterns in our data is relatively rare (less than 20% of the time for P2016+28, for example).

The drift of the features in the dynamic spectra can be explained by a refractive shift of the diffractive intensity pattern at the observing plane,[1,2,4] as illustrated by the simple schematics of Fig. 4. Here we use the thin screen model for the scattering medium. The phase function at the screen can be decomposed[1] into a rapidly varying or "diffractive" part, $\phi_d(x)$, and a relatively slowly varying "refractive" part, $\phi_r(x)$ (Fig. 4a). The diffractive phase fluctuations give rise to a diffractive intensity pattern at the observing plane, $z = z_o$. The effect of the refractive phase fluctuations is to shift the diffractive pattern by different amounts X at different positions on the observing plane (Fig 4b). This shift is given by $X = (z\lambda/2\pi)(\partial\phi_r/\partial x)$.

Here ϕ_r itself varies directly with wavelength. Therefore, the refractive shift is inversely proportional to the square of the frequency. Under these conditions, a diffractive intensity maximum at two different frequencies is shifted to different locations in the observing plane, at different frequencies as illustrated in Fig. 4c. The motion of the observer through such a intensity pattern with a relative velocity V converts the spatial separation between intensity maxima (or minima) at different frequencies to a corresponding temporal

Refractive Phenomena

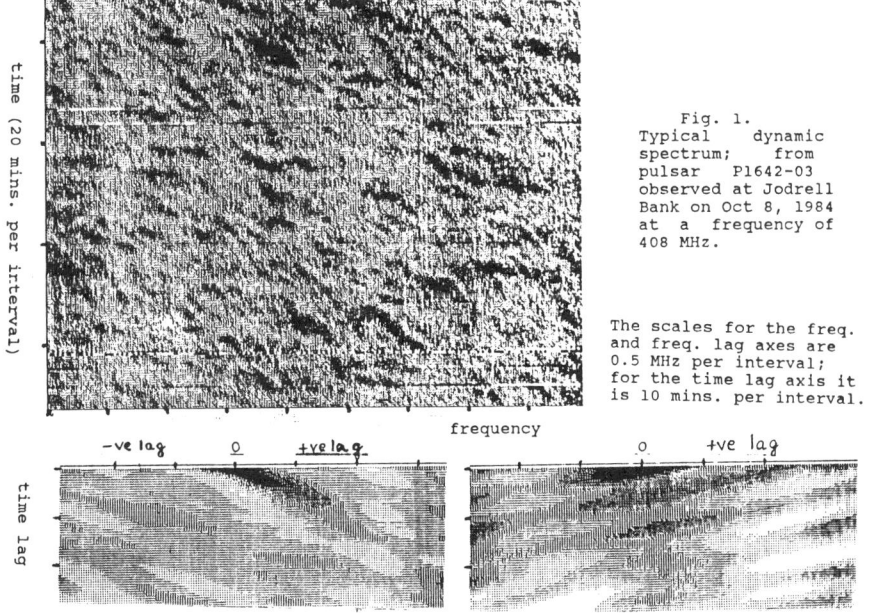

Fig. 1. Typical dynamic spectrum; from pulsar P1642-03 observed at Jodrell Bank on Oct 8, 1984 at a frequency of 408 MHz.

The scales for the freq. and freq. lag axes are 0.5 MHz per interval; for the time lag axis it is 10 mins. per interval.

Fig. 2. Correlation functions for dynamic spectra of P0823+26 at two epochs (Oct 9,1984 and Jan 28,1985) showing change in drift of features.

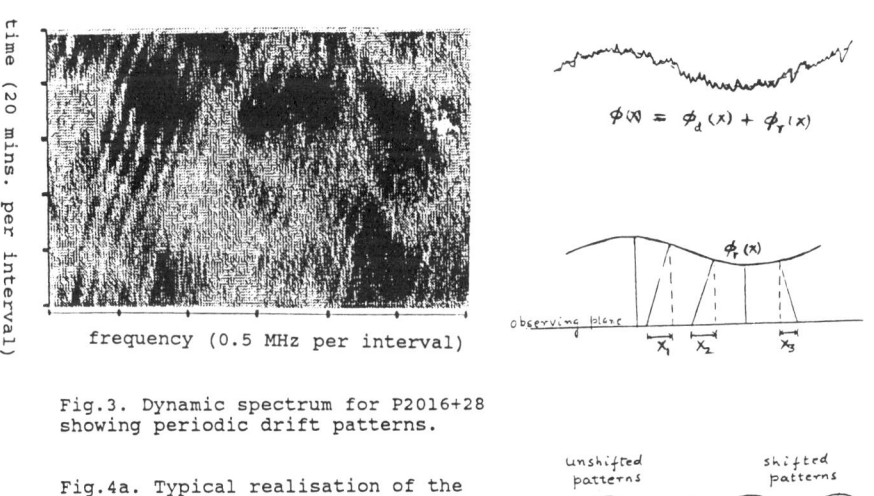

$\phi(x) = \phi_d(x) + \phi_r(x)$

Fig.3. Dynamic spectrum for P2016+28 showing periodic drift patterns.

Fig.4a. Typical realisation of the phase function at the screen.

Fig.4b. Shifting of diffractive patterns due to refractive effect.

Fig.4c. Frequency dependence of the refractive shift.

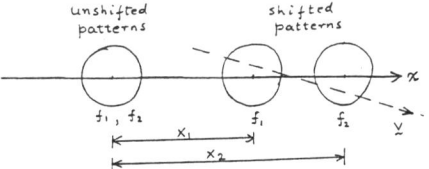

delay. This gives rise to the characteristic drift of the features in the resulting dynamic spectrum. With slow and random variations of the gradient of the refractive phase fluctuations at the screen, the refractive shift would vary in magnitude and direction and if we assume V to remain constant, we would expect to see the drift of the features change slowly and randomly with time as X fluctuates randomly.[8]

To do a quantitative analysis of the situation, we look at the two dimensional correlation function for the dynamic spectra. The correlation function for the intensity under conditions of a refractive shift $C_r(v,\tau)$ is given as

$$C_r(v,\tau) = C_d \left[V_x\tau - \frac{2xv}{f},\ V_y\tau - \frac{2yv}{f},\ v \right] \quad (1)$$

where $C_d(\xi, \eta, v)$ is the correlation function for the purely diffractive pattern, ξ and η are the spatial lags (transformed to temporal lag due to relative motion of observer w.r.t. the pattern), v is the frequency lag, τ is the temporal lag, f is the center frequency, V_x and V_y are the components of the relative velocity and (X,Y) is the refractive shift which is assumed to be unchanging with position (x,y) (this is equivalent to assuming a linear gradient for ϕ_r).

If we assume for $C_d(v, \tau)$ a gaussian-elliptical model, we get the following expression for $C_r(v, \tau)$

$$C_r(v, \tau) = \exp\left[-(c_1 v^2 + c_2 v\tau + c_3 \tau^2)\right] \quad (2)$$

where the parameters c_1, c_2 and c_3 are given by

$$c_1 = \frac{4(X^2 + Y^2)}{f^2 r_p^2} + \frac{1}{v_p^2} \ ;\ c_2 = \frac{-4(XV_x + YV_y)}{f r_p^2} \ ;\ c_3 = \frac{V_x^2 + V_y^2}{r_p^2} \ ; \quad (3)$$

Here r_p and v_p are the spatial and frequency decorrelation widths of C_d.

Due to their dependence on the refractive shift components X and Y, we expect c_1 and c_2 to fluctuate slowly and randomly with time while c_3 should remain relatively constant. This means that the characteristic width of the correlation function along the frequency lag axis and the tilt angle of the ellipse should fluctuate randomly while the width along the time lag axis remains constant. This type of behavior is shown by the pulsar P0834+06, as illustrated in Fig. 5. On fitting our correlation data with the above model, we do indeed find the parameters for some pulsars to behave in the predicted manner. However, there are a few pulsars which show fluctuations of the correlation width along the time lag axis. P2016+28 is a notable example. It is possible to have such fluctuations if the relative velocity V is not constant. The proper motion of P2016+28 is unusually small and the motion of the earth could then cause significant periodic variations.

Under the assumptions of i) a square law structure function for the phase fluctuations; and ii) the correlation of intensity equal to the square of the second moment of the field, the following is a more accurate expression[2] for $C_d(v, \tau)$

$$C_d(v, \tau) = \frac{1}{(1 + v^2/v_p^2)} \exp\left[-\frac{(V_x^2 + V_y^2)\tau^2}{r_p^2 (1 + v^2/v_p^2)} \right] \quad (4)$$

This gives $C_r(\nu, \tau)$ as

$$C_r(\nu, \tau) = \frac{1}{1 + d_4 \nu^2} \exp\left[-(d_1 \nu^2 + d_2 \nu \tau + d_3 \tau^2)/(1 + d_4 \nu^2)\right] \quad (5)$$

where the four parameters d_1, d_2, d_3 and d_4 are given by

$$d_1 = \frac{4X^2 + Y^2}{f^2 r_p^2} \; ; \; d_2 = \frac{-4(XV_x + YV_y)}{f \, r_p^2} \; ; \; d_3 = \frac{V_x^2 + V_y^2}{r_p^2} \; ; \; d_4 = \nu_p^{-2} \quad (6)$$

Once again, we expect d_1, d_2 to fluctuate slowly and randomly and d_3, d_4 to remain relatively constant. Fig. 6 shows a typical time series of these parameters for pulsar P0823+26. In general, we do not find the parameters to vary in the manner expected. We plan to pursue more sophisticated methods of including the refractive effects. The possibilities include i) effect of "curvature" in the refractive phase front[1] - in this case the refractive shift would be a function of position (x,y) also; and ii) a range of velocities[7] contributing to the scintillation pattern.

References

1. J.M. Cordes, A. Pidwerbetsky, A. and R.V.E. Lovelace, *Ap. J.*, (1986).
2. I.V. Chashei and V.I. Shishov, *Astron. Zh.*, **53,** 26 (1976).
3. J.M. Cordes, J.M. Weisberg and V. Boriakoff, *Ap. J.*, **268,** 370 (1983).
4. A. Hewish, *M.N.R.A.S.*, **192,** 799 (1980).
5. A. Hewish, A. Wolszczan and D. Graham, *M.N.R.A.S.*, **213,** 167 (1985).
6. J.A. Roberts and J.G. Ables, *M.N.R.A.S.*, **201,** 1119 (1982).
7. V.I. Shishov, *Soviet Astron.* **17,** 598 (1974).
8. F.G. Smith, and N.C. Wright, *M.N.R.A.S.*, **214,** 97 (1985).

144 Radio Wave Scattering in the Interstellar Medium

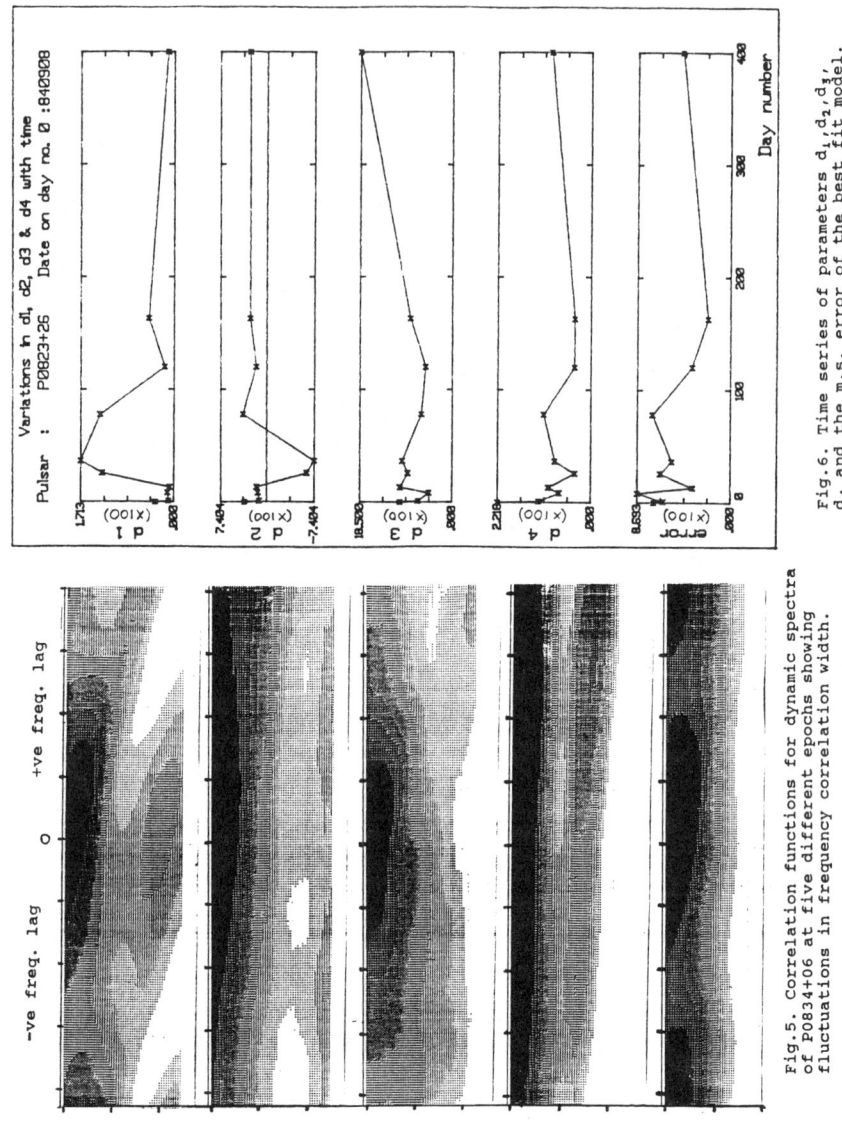

Fig.6. Time series of parameters d_1, d_2, d_3, d_4 and the m.s. error of the best fit model.

Fig.5. Correlation functions for dynamic spectra of P0834+06 at five different epochs showing fluctuations in frequency correlation width.

Refractive Scintillation, Caustics, and Interstellar
Interferometry of the Pulsar PSR 1133+16

A. Wolszczan
National Astronomy and Ionosphere Center, Arecibo

J.E. Bartlett
Carleton College

J.M. Cordes
Astronomy Department, Cornell University

ABSTRACT

A multiple imaging episode of the pulsar PSR 1133+16 has been detected with the Arecibo telescope at 1400 MHz. The episode is characterized by intensity variations by a factor of 10 on a timescale of 4 minutes. The observed pulsar 'light curve' has the form of periodic fringes resembling the oscillations expected from simple caustic events. The fringes exhibit pulse longitude dependent shifts in time and frequency similar to those recently found in PSR 1237+25. This suggests another case of successful interstellar interferometry of the pulsar magnetosphere.

INTRODUCTION

Multiple imaging effects caused by refraction of pulsar radiation in the interstellar medium (e.g. Romani, Blandford and Narayan 1986; Cordes, Pidwerbetsky and Lovelace 1986) lead to observable periodic modulations of pulsar intensity in time and frequency (Hewish, Wolszczan and Graham 1985; Cordes and Wolszczan 1986). Episodes involving only a few images are expected to generate simple, well defined oscillation patterns in the pulsar dynamic spectra (intensity as a function of time and frequency). Cordes, Pidwerbetsky and Lovelace (1986) have shown that oscillations caused by beating between two images are equivalent to fringes of an interferometer with a baseline of the order of ~ 1 A.U. and with a sub-microarcsecond resolution. This 'interstellar' interferometer has been used by Wolszczan and Cordes (1987) to resolve an apparent $\sim 10^8$ cm separation of emission regions in the magnetosphere of the pulsar PSR 1237+25. Here, we report another case of pulsar interferometry involving a double imaging episode detected during the interstellar scintillation observations of PSR 1133+16.

OBSERVATIONS

Observations of the dynamic spectra of interstellar scintillation (ISS) of PSR 1133+16 were made on 27 June 1987 with the 305-m Arecibo antenna and the 40 MHz, 3-level spectrometer at 1397 MHz. The spectra of the signals of both circular polarizations were obtained in 256 pulse longitude bins (360° of longitude equals one pulsar period) across the pulsar period. Each 40 MHz, 512-channel spectrum was integrated for 20 s, synchronously with a Doppler-corrected pulsar period. The total duration of the measurement was 34 minutes. With an effective longitude resolution of 1°.4, we obtained eleven 'on-pulse' dynamic spectra, adequately resolving the two-component pulse shape. The 'off-pulse' spectra were used for bandpass and flux calibration. Details of the data acquisition technique can be found in Wolszczan and Cordes (1987).

© 1988 American Institute of Physics

ANALYSIS OF THE DYNAMIC SPECTRA

The dynamic spectrum of PSR 1133+16 integrated over all 'on-pulse' longitude bins is shown in Figure 1. During the first 20 minutes the intensity maxima exhibit a fast drift towards lower frequencies. Later, the sense of frequency drifting changes abruptly (within 2 minutes) and this is accompanied by the increase of total pulsar flux from 0.5 Jy to about 12 Jy. The average

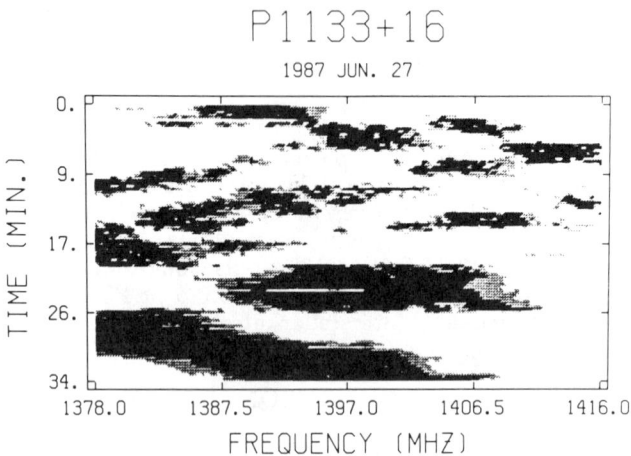

Fig. 1. Dynamic spectrum of the pulsar PSR 1133+16 at 1397 MHz.

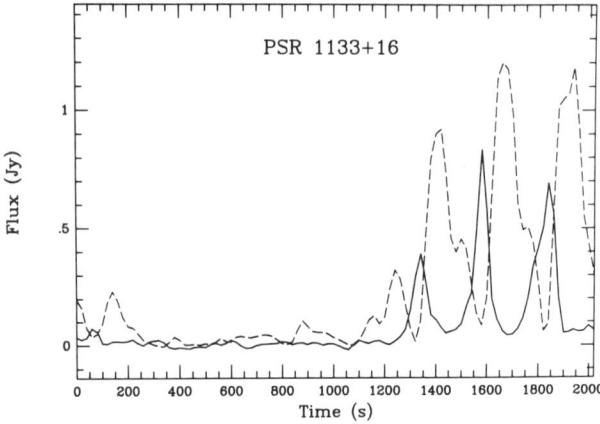

Fig. 2. Total intensity variations of PSR 1133+16 at the leading edge (*solid line*) and the trailing edge (*dashed line*) of the pulse profile.

decorrelation bandwidth, $\Delta\nu$, and the decorrelation time, Δt, of the intensity fluctuations in the dynamic spectrum are 25±5 MHz and 6±1 min., respectively. A longitude resolved analysis of these data has shown that during the period of dramatic brightening, the pulsar intensity oscillates at a period of 260 ± 10 s and the phase of intensity maxima is a function of pulse longitude. This is illustrated in Figure 2, which shows time variations of pulsar intensity at two opposite edges of the pulse profile. A similar but much less pronounced effect has been found in the frequency domain.

The pulse longitude dependent shift of intensity fringes observed in PSR 1133+16 is similar to that recently found in the dynamic spectra of PSR 1237+25 by Wolszczan and Cordes (1987) and conforms to the interpretation of intensity fringes made by Cordes, Pidwerbetsky, and Lovelace (1986). We have not detected any significant dependence of the fringe phase on the sense of circular polarization. Consequently, instrumental polarization effects are unlikely to produce fringe shifts of the observed magnitude ($\sim$ 60 s) across the pulse profile. Our analysis of the 1400 MHz spectra of PSR 0834+06, which frequently undergoes episodes of refractive multiple imaging, has not detected any measurable shifts of the observed fringe phases. Finally, the 45 s intensity modulation of PSR 1133+16 thought to be intrinsic to the pulsar (Backer 1973) has a much shorter period than observed in our data. These arguments and an apparent correlation of the pulsar brightening with morphological changes in the dynamic spectrum lead us to believe that we have in fact detected another multiple imaging event, which can be used to resolve the pulsar magnetosphere.

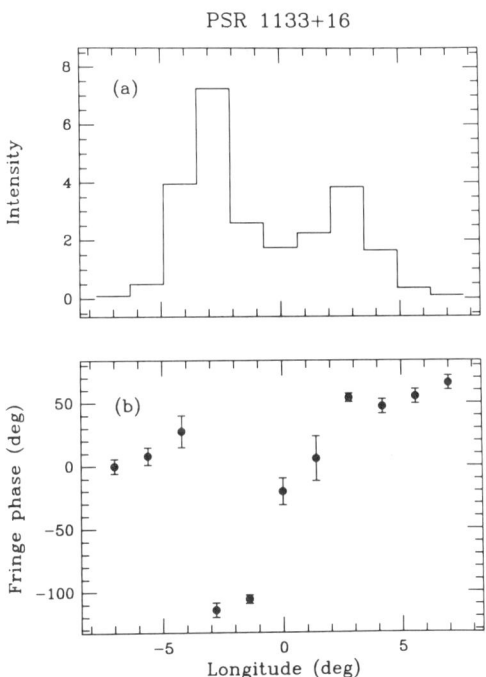

Fig. 3. (a) Pulse profile of PSR 1133+16 with 1°.4 resolution. (b) Intensity fringe phases as a function of pulse longitude.

We have computed the fringe phases at eleven 'on–pulse' longitudes from two dimensional cross–correlation functions of the reference dynamic spectrum at the leading edge of a pulse profile with the dynamic spectra at all other longitudes. The phases, expressed in degrees (here, 360° of phase corresponds to one intensity oscillation period, $P_t = 260$ s) are shown in Figure 3 together with the pulse profile.

DISCUSSION

Our results can be satisfactorily described in terms of a simple thin screen model of refraction in the interstellar medium, as shown in Cordes and Wolszczan (1988; these proceedings). In this model, the temporal intensity oscillation in the dynamic spectrum has a period equal to the reciprocal of the dot product of the pulsar proper motion velocity and the transverse gradient of the phase difference (eqn [2] of Cordes and Wolszczan (1988)). The phase change associated with a transverse displacement of the source δx_s involves the same gradient, and therefore δx_s is simply

$$\delta x_s = V_\perp \delta t, \tag{1}$$

where δt is the measured fringe phase shift in time. In eqn (1) we have ignored factors of order unity that involve orientations of the proper motion, phase gradient, and source position vectors (see eqn [3] of Cordes and Wolszczan (1988)). Given the values of $V_\perp =264$ km/s (Lyne, Anderson and Salter 1982), and $\delta t_{max}=83$ s, eqn (1) yields a maximum observed transverse separation between the emission regions of PSR 1133+16, $x_s \sim 3 \times 10^9$ cm. This value is similar to the separations deduced from the measured fringe shifts in frequency during a double imaging event of PSR 1237+25 (Wolszczan and Cordes 1987).

The linearly polarized emission of PSR 1133+16 exhibits the well known phenomenon of orthogonal polarization modes (Backer and Rankin 1980). The presence of polarization modes is particularly well visible at the longitudes of component I of the pulse profile and has the form of a 90° discontinuity in the polarization angle curve as a function of pulse longitude (e.g. Stinebring 1982). This correlates very well with a longitude dependence of intensity fringe phases shown in Figure 3. In principle, such correlation is not surprising, since the longitude variation of both the polarization angle and the fringe phase is determined by geometry of the pulsar emission (Cordes and Wolszczan 1988, and references therein).

The intensity oscillation observed in PSR 1133+16 is similar to an intensity pattern of a simple fold caustic event, caused by beating between two merging subimages (Berry and Upstill 1980). An occurence of such events under the conditions of strong scattering in the interstellar medium has been considered by Goodman et al.(1987). Since PSR 1133+16 is a nearby, low dispersion measure pulsar ($DM = 4.85\ pc \cdot cm^{-3}$), it may already be in the weak scattering regime, when observed at 1400 MHz (e.g. Backer 1975). When scintillation is weak, a corresponding diffraction scale size is greater than the Fresnel scale, $r_F = \sqrt{\lambda z}$ (λ is the wavelength and z is the pulsar distance) and a scattered pulsar image breaks up into a few subimages that rearrange themselves on a timescale $t \sim r_F/V_\perp$. In our case, with $z \sim 160$ pc and $\lambda = 21$ cm, we obtain $t \sim 1$ hour, which is not inconsistent with a lower limit of ~ 15 min to the duration of the observed event.

The results presented here, together with a double imaging event of PSR 1237+25 described by Wolszczan and Cordes (1987), demonstrate that the occurence of interstellar interferometry conditions in the refracting interstellar medium may be frequent enough to use it as a practical tool to study pulsar magnetospheres. This attractive possibility and two-dimensional models of the refracting screen and the pulsar magnetosphere are further explored by Cordes and Wolszczan (1988).

This work was supported by the National Astronomy and Ionosphere Center, which operates the Arecibo Observatory under contract to the National Science Foundation.

REFERENCES

1. D.C. Backer, *Ap. J.*, **182**, 245 (1973).
2. D.C. Backer, *Astr. Ap.*, **43**, 395 (1975).
3. D.C. Backer and J.M. Rankin, *Ap. J. Suppl.*, **42**, 143 (1980).
4. M.V. Berry and C. Upstill, *Progress in Optics*,**18**, ed E. Wolf (Amsterdam, North–Holland), 257 (1980).
5. J.M. Cordes, A. Pidwerbetsky and R.V.E. Lovelace, *Ap. J.*, **310**, 737 (1986).
6. J.M. Cordes and A. Wolszczan, *Ap. J. (Letters)*, **307**, L27 (1986).
7. J.M. Cordes and A. Wolszczan, these proceedings (1988).
8. J. Goodman, R.W. Romani, R. Blandford and R. Narayan, *M.N.R.A.S.*, **229**, 73 (1987).
9. A. Hewish, A. Wolszczan and D.A. Graham, *M.N.R.A.S.*, **213**, 167 (19 85).
10. A.G. Lyne, B. Anderson and C.J. Salter, *M.N.R.A.S.*, **201**, 503 (198 2).
11. R.W. Romani, R. Narayan and R. Blandford, *M.N.R.A.S.*, **220**, 19 (19 86).
12. D.R. Stinebring, *Ph. D. Thesis*, Cornell University (1982).
13. A. Wolszczan and J.M. Cordes, *Ap. J. (Letters)*, **320**, L35 (1987).

EXTREME SCATTERING EVENTS

R. Fiedler, R. Simon, and K. Johnston
Naval Research Laboratory, Washington, D.C. 20375

B. Dennison
VPI&SU, Blacksburg, VA 24061

A. Hewish
Mullard Radio Astronomy Observatory, Cambridge CB3 0HE, UK

ABSTRACT

Daily flux density measurements of compact extragalactic radio sources using the Green Bank interferometer (36 sources)[1] and the NRAO 300ft antenna (340 sources)[2] reveal several unusual variations in the light curves that do not follow typical source variations. Unusual variations have been identified in the light curves of eight quasars. Refractive effects involving small-scale inhomogeneities in an ionized structure in the interstellar medium seem the most likely explanation for the unusual variations observed.

INTRODUCTION

The data and discussion presented in this paper are based on the 9 year Green Bank interferometer monitoring program and some preliminary results from the 6 month NRAO 300 ft antenna monitoring program.

Our observations suggest that during the course of monitoring numerous QSO's, compact ionized structures in the interstellar medium drifted through the line-of-sight to some of those QSO's, resulting in the unusual one to three month long variations in their light curves.[3] These structures, for reasons of pressure equilibrium discussed below, are probably extended in one or two dimensions rather than spherical in shape. Whether these structures are associated with shocks, filaments, or some other perturbation to the interstellar medium is unclear.

That these variations have not been clearly identified before attests to the importance of frequent monitoring, at least twice weekly. The rapid excursions in flux density span one to two weeks, whereas the minima usually associated with these events last one to two months. Unusually high or low flux density points observed in other monitoring programs could well be the result of extreme scattering events.

The intent of the 6 month monitoring program using the NRAO 300ft antenna is to trigger VLBI, VLA, and optical observations to better estimate the physical conditions of the ionized structures assumed to cause the unusual flux density variations. We hope that

© 1988 American Institute of Physics

multifrequency observations taken during and after an event will
better define this new astrophysical phenomenon.

OBSERVATIONS

The Green Bank interferometer was used to measure the flux
densities of 36 QSO's on a daily basis during the period January
1979 to January 1988. The frequencies of observation were 2695 and
8085 MHz. The interferometer was operated during this period by
the National Radio Astronomy Observatory (NRAO) for the U.S. Naval
Observatory (USNO) as part of a program to determine earth rotation
parameters.[2] The Naval Research Laboratory (NRL) used the same
data to study compact extragalactic radio source variability.[1]

In January 1988 NRL began using the interferometer solely as a
monitoring instrument for compact extragalactic source variability,
observing approximately 100 sources every other day.

A shorter term monitoring program lasting six months began in
December 1987 using the NRAO 300ft antenna. Approximately 340
sources were observed every other day at 3265 and 4675 MHz.

UNUSUAL VARIATIONS

Unprecedented variations have been observed in the gigahertz
frequency light curves of eight quasars: 0300+470; 0333+321;
0954+658; 1502+106; 1611+343; 1749+096; 1821+107; and 2352+495.
The most dramatic variations occurred in 0954+658. See figure 1.
This is the only source in our list that shows unusual variability
at 8085 MHz during the lower frequency 2695 MHz event. Common to
all of these sources, however, is the superposition of time scales
ranging from 10 days to 3 months.

Only one other source in our list 1502+106 shows a clear
signature of paired maxima surrounding a prolonged minimum as in
0954+658. Four of the sources 0300+470, 1611+343, 1741+096, and
2352+495 show oscillatory events much like a damped sinusoid. Two
of the sources 0333+321 and 1821+107 have events that look like
inverted gaussians, except that the light curve of 0333+321
contains two events separated by 1.25 years. We emphasize that in
all of these light curves, except for 0954+658, the variations are
apparent only at the lower frequency.

These light curves and the polarization data from the NRAO 300ft
antenna observing program are discussed in detail by Fiedler et
al..[4]

DISCUSSION

It is apparent from figure 1 that the variations we call an
event are atypical. This observation is further supported by a
qualitative comparison of an event with the common understanding of
intrinsic variability and normal interstellar scintillation.

The sudden release of energy by a compact extragalactic object and subsequent propagation in the form of electromagnetic radiation through intervening material to the observer are in this case identified with effects intrinsic to the source and the interstellar medium, respectively. Propagation effects, such as refraction and diffraction, modulate intrinsic variability.

Normal intrinsic variability is theoretically identified with an injection of an ensemble of relativistic particles that subsequently expands into the surrounding medium, resulting in a shift in the synchrotron spectrum to lower frequencies and flux densities. At gigahertz frequencies, whether the emitted radiation represents cooling from an expanding cloud[5,6,7] or a shock front[8,9], the lower frequency variations appear to be attenuated, smoothed, and time delayed with respect to the higher frequency variations.

Notice in figure 1 that, excluding the event, the 2.7 GHz variations lag those at 8.1 GHz by about 56 days. Conversely, the event at 8.1 GHz appears symmetrically aligned within the 2.7 GHz event and that the variations at both frequencies bear no resemblance to each other.

Other intrinsic effects could include motion within the quasar itself. In reference 1 we derived a proper motion of 0.1 mas/day. This implies a linear motion at 0954+658 of 500c. Since this is more than an order of magnitude in excess of velocities inferred for superluminal VLBI components, explanations involving motion within the quasar, or at cosmological distances, seem unlikely. The speed-of-light distance is approximately 2 Mpc.

The unusual variations illustrated in figure 1 are not likely to be identified with intrinsic variability, or with normal interstellar scintillation. Gigahertz frequency scintillation is generally persistent over the entire period of observation with time scales that range from days to years depending on the galactic latitude of the extragalactic object and its intrinsic angular size. The rms modulation amplitude of the light curve produced by scintillation is typically a few percent of the mean flux density of the source and is stronger at 2.7 GHz than at 8.1 GHz.

Evidently, the ionized structures represent a new astrophysical phenomenon. Assuming the ionized structure is within our own galaxy having a velocity of v < 200 km/s, the observed proper motion implies a distance of d < 1.3 kpc. Furthermore, the 60 day minimum in the 2.7 GHz light curve of 0954+658 implies a projected shadow at the earth of less than 7 AU. This should be comparable to the physical size of the ionized structure.

Two models have been proposed to account for the details of the variations in the radio light curves. One involves refractive scattering[3] and the other uses caustic surfaces[10,11]. The refractive scattering model requires scattering through an ensemble of irregularities in ionized gas density. The strong 8.1 GHz spikes in 0954+658 require that the irregularities must focus strongly with occasional multiple imaging. Although, to observe these effects at 8.1 GHz the observer must be in the focal plane and pass close to the cloud-QSO line. At 2.7 GHz the refraction angle is about 9 times larger resulting in a local minimum where radiation is scattered away from the line-of-sight through the center of the ionized structure. Paired maxima surrounding the prolonged minimum are a consequence of the normal intensity augmented by the scattered radiation. For an estimated refraction angle of 0.5 mas at 8.1 GHz the column density is estimated to be less than 10^{18} cm^{-2}.

Alternatively, the observed events may be due to an observer crossing caustic surfaces caused by a single density inhomogeneity in the ISM. The 4 spikes at 8.1 GHz are interpreted as the crossing of 4 fold lines. At 2.7 GHz the increased refraction angle separates the caustics further, and when convolved with an increased source size results in two broad peaks.

We now address the geometry of the cloud. A spherical cloud with diameter 7 AU would require an ionized particle density of roughly 6 x 10^3 cm^{-3}. This particle density at a temperature of a few thousand degrees Kelvin will put this hypothetical cloud in overpressure with the surrounding medium by three to four orders of magnitude. However, if the ionized structure were filamentary or planar, as suggested by Romani, Blandford, and Cordes[10], magnetic confinement may bring the 'cloud' into equilibrium with the surrounding interstellar medium.

We conclude by estimating an upper limit to the covering factor of ionized structures on the sky. Eight events spanning 4.9 years in 275 source-years corresponds to a given point source being affected about 2% of the time. Assuming a galactic disk population of ionized structures, this implies a number density of < 240 pc^{-3} h^{-1}, where h is the ratio of the linear extent to the width of the typical filament. This number density is more than three orders of magnitude greater than that for stars in the galaxy. Integrated over the entire galaxy, the ionized structures would account for about 100 solar masses.

CONCLUSIONS

We are able to infer from the light curve of 0954+658 the following physical parameters for these ionized structure in the interstellar medium.

1. transverse dimension < 7 AU
2. number density < 240 pc^{-3}
3. column density < 10^{18} cm^{-2}

In addition, hydrostatic equilibrium suggests that the ionized structures are filamentary rather than spherical in shape.

More importantly, we find that there appears to be a large variety in the signature of an extreme scattering event in radio light curves. Simple inverted gaussians, paired maxima surrounding a prolonged minimum, oscillations, and repeated events in the same source all suggest that the interstellar medium on the length scales that we seem to be sampling is very complicated.[4] This should not be surprising, but will remain speculation until direct observations of an ionized structure are performed.

REFERENCES

1. R.L. Fiedler, E.B. Waltman, J.H. Spencer, K.J. Johnston, P.E. Angerhofer, D.R. Florkowski, F.J. Josties, W.J. Klepczynski, D.D. McCarthy, and D.N. Matsakis, Ap. J. Supp. 65, 319 (1987).

2. D.N. Matsakis, F.J. Josties, P.E. Angerhofer, D.R. Florkowski, D.D. McCarthy, X. Jiayan, and P. Yunlou, Astron. J. 91, 1463 (1986).

3. R.L. Fiedler, B. Dennison, K.J. Johnston, and A. Hewish, Nature 326, 675 (1987).

4. R.L. Fiedler, B. Dennison, R. Simon, and K.J. Johnston, in progress.

5. Van der Laan, Nature 211, 1131 (1966).

6. K.I. Kellermann, and I.I.K. Pauliny-Toth, Ann. Rev. Astron. Ap. 19, 373 (1981).

7. C.P. O'Dea, W.A. Dent, and T.J. Balonek, T.J., in Proceedings of the Manchester Meeting on Active Galactic Nuclei, ed. J.E. Dyson (Manchester: Manchester University Press), (1984).

8. A.P. Marscher and W.K. Gear, Ap.J. 298, 114 (1985).

9. P.A. Hughes, H.D. Aller, and M.F. Aller, 1985, Ap.J. 298, 301 (1985).

10. R.W. Romani, R.D. Blandford, and J.M. Cordes, Nature 328, 324 (1987).

11. J.J. Goodman, R.W. Romani, R.D. Blandford, and R. Narayan, M.N.R.A.S. 229, 73 (1987).

Refractive Phenomena

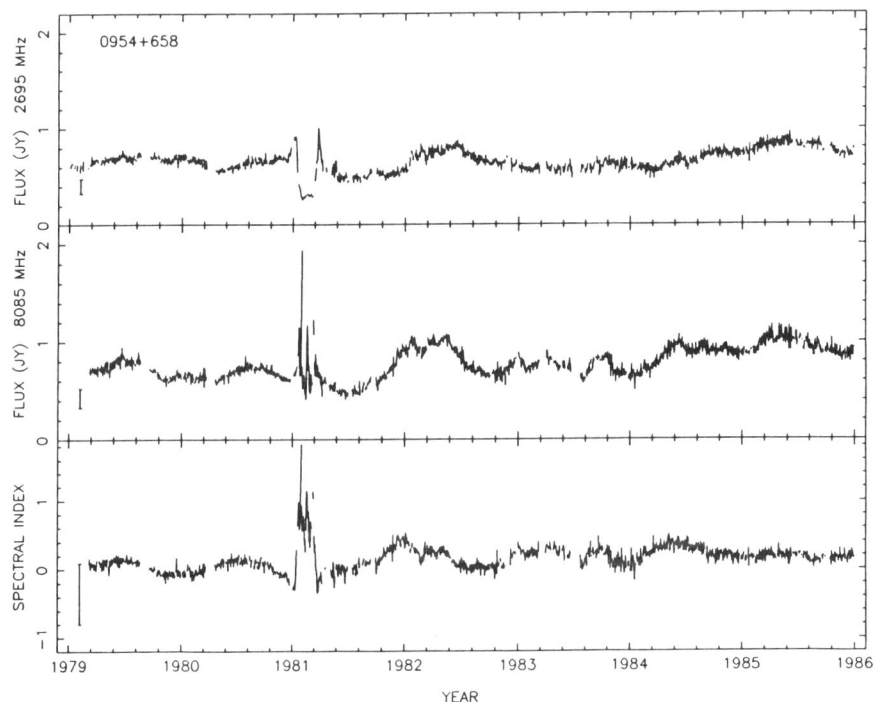

Figure 1. The radio light curve of the quasar 0954+658 at 2695 and 8085 MHz. The unusual variations during 1980.95-1981.30 are probably due to an extreme scattering event.

PROPERTIES OF LARGE DIM REFRACTORS

Roger W. Romani

Department of Astronomy, University of California,
Berkeley, CA 94720

ABSTRACT

We discuss the dramatic refraction events recently noticed in the 2.7 GHz light curves of several compact extragalactic radio sources[1], describing physical conditions in the refractors and suggesting an association with aging supernova remnants. The proposed properties are constrained by a number of extant observations; future studies which should further elucidate conditions in these dense, ionized structures are proposed.

INTRODUCTION

The remarkable light curve fluctuations discovered in Green Bank Interferometer observations of several extragalactic radio sources by Fiedler et al.[1], if interpreted as a propagation effect, provide important insights into the structure of the ionized interstellar medium (ISM) at AU lengthscales[2]. These fluctuations seem to have a common signature: at 2.7 GHz there is a substantial minimum in the flux, bracketed by two maxima with 1.5-2 times the mean flux. The whole event lasts $\sim 10^7 \tau_7$ seconds and a crude estimate suggests that the duty cycle for such events is quite large, $f \gtrsim 0.005$. Several scattering models have been proposed for the particularly prominent event seen at both 2.7 GHz and 8.1 GHz in 0954+658[1-3], but we concentrate here on the more 'typical' 2.7 GHz modulations. Taking the two parameters above as characteristic, one finds that even these events have inferred electron density fluctuations much larger than anticipated from the general scattering medium. We note that consideration of pressure and ionization place substantial constraints on, and suggest sites for this extensive, high-density, highly clumped phase of the ISM called here the Dense Ionized Medium or DIM.

II. BASIC PROPERTIES

We adopt the view that the fluctuations observed by Fiedler et al. are due to strong focusing (i.e. ray crossing and the formation of caustics) by the galactic interstellar medium. Then, since the rays cross $\alpha \sim 1$ times between the refractor and the observer at $\nu = 2.7$ GHz, the observed event duration τ, gives immediately the electron column density

$$N_e = \frac{2\pi\alpha\nu^2(\tau v)^2}{r_e c^2 D} \sim 5 \times 10^{17} (\tau_7 v_7)^2 / D_{kpc} \text{ cm}^{-2} \qquad (1)$$

and the fluctuation size $a \sim 10^{14} \tau_7 v_7$ cm for a distance D_{kpc} kpc. Although the duration of the event gives an estimate for the transverse size of the refractor, the line-of-sight dimension might be substantially larger if the fluctuations are filamentary or sheet-like. Adopting a large, but not unreasonable, value for this elongation, $\eta = 10^2 \eta_2$ (with $\eta_2 \sim 1$), provides a minimal estimate for the electron density $n_e \sim N_e/(\eta_2 a) \sim 50$ cm^{-3}, and thus gives the emission measure associated with the fluctuation as

$$EM = \eta a n_e^2 = \alpha^2 \left(\frac{2\pi\nu^2}{r_e c^2}\right)^2 \frac{(\tau v)^3}{D^2} \eta^{-1} \sim 10 \frac{(\tau_7 v_7)^3}{D_{kpc}^2} \eta_2^{-1} \text{ cm}^{-6}\text{pc}. \quad (2)$$

If the strong modulation caused by the caustic is to be observed, the angular size of the background source must be a factor of a few smaller than that of the refractor. Extragalactic sources limited by self absorption will have intrinsic angular sizes $\theta_I \sim 0.5(2.7\text{GHz}/\nu)$ mas; scattering from the distributed electron density fluctuations in the galactic disk will give other sources (*e.g.* pulsars) a minimum angular size[4] $\theta_s \sim 0.02 C_{-4}^{1/2}(2.7\text{GHz}/\nu)^2$ mas, where at high latitude $C_{-4} \sim 10$. This implies that the lensing fluctuation has a maximum distance $D_{kpc} \lesssim 3\tau_7 v_7$ ($D_{kpc} \lesssim 100\tau_7 v_7 C_{-4}^{-1/2}$ when scattering dominates) and that a lower bound on the emission measure is $EM \gtrsim 0.5\tau_7 v_7 \eta_2^{-1}$cm^{-6}pc.

The small angular size $\sim a/D_{kpc}$ of these perturbations implies substantial beam dilution for any optical observation and small emission measure fluctuations associated with particular events. However, the duty cycle f above (which gives an estimate of the mean covering fraction on the sky) is relatively large so the net emission from the population of scatterers can be substantial. We suppose that the caustic-producing refractors are sheets of ionized gas, whose facets have have a typical thickness a and transverse size ηa. In ref. 2 it was suggested that such structures might be present in old supernova remnants; this site is considered further below. With this choice of geometry, η^{-2} of the sheets are viewed edge-on so the mean additional dispersion measure in a region with a duty cycle for caustic events $f = 10^{-2} f_{-2}$ (with $f_{-2} \sim 1$) is

$$\overline{DM} = \eta^2(N_e/\eta)f \\ \sim 0.2((\tau_7 v_7)^2/D_{kpc})\eta_2 f_{-2} \text{ cm}^{-3}\text{pc} \gtrsim 0.04(\tau_7 v_7)\eta_2 f_{-2} \text{ cm}^{-3}\text{pc} \quad (3)$$

and the corresponding mean emission measure is

$$\overline{EM} = \eta^2(EM/\eta)f \\ \sim 10(\tau_7 v_7)^3/D_{kpc}^2 f_{-2} \text{ cm}^{-6}\text{pc} \gtrsim 0.5(\tau_7 v_7)f_{-2} \text{ cm}^{-6}\text{pc}, \quad (4)$$

where the lower bounds come from the finite intrinsic source size at 2.7 GHz above.

Clearly, the contribution to the electron column through the galactic disk (3) is quite negligible even for lines of sight at high galactic latitude, as pulsar dispersion measures[5] give a lower limit of $DM_\perp \gtrsim 30 \text{cm}^{-3}\text{pc}$. The high density implied for the refractors is, however, reflected in a rather large mean EM. Observations of the the diffuse galactic $H\alpha$ emission[6] give an emission measure in low velocity ($\lesssim 50 \text{km/s}$) gas of $EM \sim 2.8 T_4^{0.9} \text{cm}^{-6}\text{pc}$, where the gas temperature is $10^4 T_4 \text{K}$. Direct measurements of the electrons from low frequency ($\sim$MHz) free-free absorption of the diffuse galactic background[7] suggest similar values: $EM \gtrsim 4 T_4^{-1.4} \text{cm}^{-6}\text{pc}$. The emission measure for the caustic-producing refractors above then contributes at least $\sim 10\%$ of, and may in fact dominate, the diffuse galactic emission. So the DIM fluctuations are, in fact rather bright, and correspondingly, the energy demanded to keep this medium ionized can be substantial: $dE/dt \sim 2.5 \times 10^{41} \overline{EM} T_4^{-0.9} \text{erg s}^{-1}$ if our local estimates apply to $\pi(15\text{kpc})^2$ of galactic disk. For comparison, the flux of ionizing photons from O stars is $\sim 1 - 3 \times 10^{42} \text{erg s}^{-1}$ and supernovae at $(10^{51}\text{erg}/30 \text{ yr})$ contribute $\sim 10^{41} \text{erg s}^{-1}$. Apparently, the energy budget for even a modest total EM is large enough to severely restrict our sites for the DIM refractors.

III. DIM SITES and PHYSICAL CONDITIONS

With the number densities for the lensing sheets given above, and ionizing temperatures $T_4 \sim 1$, the typical caustic refractor has an associated pressure $\sim 5 \times 10^5 \tau_7 v_7 T_4 / (D_{kpc} \eta_2) \text{cm}^{-3}\text{K}$. This is ~ 100 times the mean interstellar pressure and the ionized regions would be ephemeral. A plausible site for such pressures and temperatures can be found in old supernova remnants of diameter $\lesssim 100 R_2 \text{pc}$, which would have a mean pressure $\bar{p} \sim 5 \times 10^5 (E_{51}/R_2^3) \text{cm}^{-3}\text{K}$. In figure 1, the positions of the 33 sources in the Green Bank survey[1] are shown along with small circles showing the rough positions of galactic radio continuum loops[8]; sources showing probable and possible DIM-refractor induced modulations are indicated. While not all of the 33 sources will be compact enough to show strong focusing, the events seen appear correlated with the well-know radio loops I and III. Of seven suggested events, five are within a few degrees of a loop and in particular 0954+658 is quite close to the radio ridge of loop III. Since the loops are large and irregular, significance is difficult to estimate (crudely, 10° bands around the loops covers ~ 0.2 of the sky, giving a chance association $\lesssim 0.01$). However, higher resolution radio continuum maps suggest a better correlation and the other two sources at $(l=60, b=45)$ are quite near a more distant loop visible in high-velocity HI emission[9]. Such structures have also been proposed as preferred sites for refractive low-frequency variability and other forms of scintillation.[10–12]

The radio continuum loops are believed to be old (several $\times 10^6 \text{y}$) remnants of either energetic supernovae or, more likely, the collective superbubbles blown by several closely spaced supernovae from OB associations[13]. The distances to the loops are estimated as $\sim 100 - 300$ pc and the expansion velocities have been estimated from associated HI gas as $\lesssim 50 \text{ km s}^{-1}$. In some cases, then, Earth orbital motion ($v_7 \lesssim 0.3$) will dominate the line-of-sight velocities through the

DIM fluctuations and N_e, EM and DM should be close to the fiducial values estimated above. Since the column density (1) is close to the cooling column for a modest velocity shock[14] ($v_s \sim 30$ km s^{-1}) it is appealing to associate the electron density structures with post-shock gas at the edge of the expanding loops.

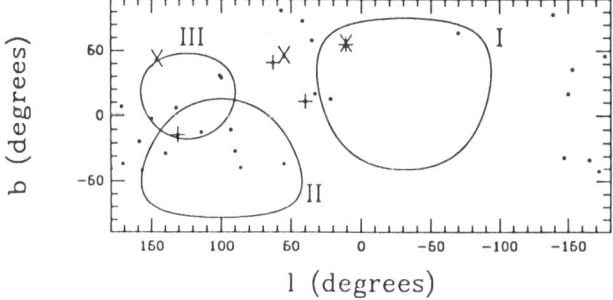

Figure 1. Sources monitored at Green Bank. Those with probable (×) and possible (+) refraction events are indicated, along with three radio continuum loops.

Note that the typical observed expansion velocities ~ 30 km s^{-1} will not provide sufficient energy to fully ionize the gas. However, sequential shocks from supernovae separated by of order the cooling time can have the desired cumulative effect. Early energetic shocks can ionize and, after modest cooling, produced shocked regions with densities several times that of the preshock gas (which might be warm interstellar medium at ~ 0.3 cm^{-3}). A subsequent shock can then produce further compression and ionized fluctuations of the appropriate column density. However for preshock densities of a few the cooling length will be $\gtrsim 10^{16}$ cm; thus in order to see refractive fluctuations on the observed timescales it is then necessary to posit that these shocks have additional structure modulating the electron density on scales of a few AU. Moreover, to produce the isolated light curve modulations seen, it is necessary that these electron column fluctuations be both strong (contrast $\gtrsim 1$) and isolated (separated by several $a \sim \tau v$). To preserve the association with a cooling column, the modulation could be transverse to the shock front, perhaps effected by ordered magnetic fields or by fluctuations associated with the Larmour radii of energetic ($\sim 10^4$ GeV), parallel propagating particles. To show how isolated pairs of caustic maxima can form in this geometry, we plot in figure 2a) the projected column from a gently curved ionized slab with an AU-scale sinusoidal modulation, along with the corresponding light curve with refractor-induced focusing. Perhaps more naturally, small scale structure can form parallel to shock fronts through cooling-instability driven amplification of preexisting density fluctuations[15]. Here much of the post-shock ionized gas may collapse into a few thin sheets with a maximum density enhancement e, and a minimum width

$\sim L_{cool}/e$, limited by either thermal conduction ($L_e \gtrsim 10^{-2} L_{cool}$ for relevant densities and pressures) or magnetic pressure ($n_{max} \sim 80(n_0^{3/2} v_7/B_{0,-6})\text{cm}^{-3}$; with pre-shock values $n_0 \sim 1$, $B_0 \sim 10^{-6}\text{G}$)[9]. These limiting densities and compressions are comparable to those inferred from the observations; the weak preexisting fluctuations on scales $\sim \eta a$ can be the *residua* of prior ionizing shocks. In figure 2b), we show schematically the density profile arising in such a shock, viewed edge-on along with the associated refractive intensity fluctuations.

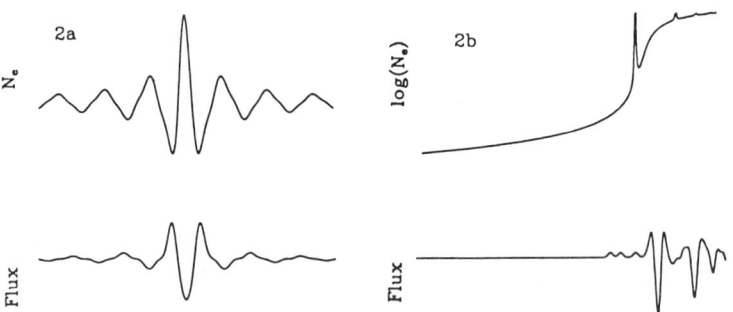

Figure 2. Above: column density profiles of (a) a curved corrugated sheet, (b) a shock with cooling substructure. Below: the corresponding light curves showing refractive 'caustic' events.

IV. ANCILLARY AND FUTURE OBSERVATIONS

The presence of electron-density perturbations on these scales, while highlighted by the caustic events seen by Fiedler *et al.* is not unanticipated in view of related observations. In particular, periodicities observed in pulsar dynamic scintillation spectra[17] and the excess low-frequency variability of pulsars and compact extragalactic sources seen at meter wavelengths suggest that refraction over $\sim$AU scales is often substantially in excess of that expected from extrapolation of the density fluctuation spectra seen at smaller scales. The refractors responsible may be weaker versions of the perturbations invoked above, or might be the same objects viewed at average angle of inclination, rather than edge-on.

The importance of magnetic fields in the radio loops is suggested by their prominence in synchrotron continuum emission, rotation measure, and optical polarization maps. If, as suggested, the structure and maximum compression of the post-shock gas is controlled by these fields, then such maps should have substantial features at milliarcsecond scales. In particular, since a magnetic field of $\sim 10^{-5} B_{-5}$ G and the column density of a typical DIM refractor will produce a rotation measure of $\sim 1.3 B_{-5}$ rad m^{-2}, phase angle fluctuations caused by individual events will be measurable at GHz frequencies. In fact, a rotation measure variation of ~ 2 rad m^{-2} over $\sim 10^3$ days with an associated DM fluctuation of ~ 0.2 cm^{-3}pc, values reasonably close to those suggested above, have been seen for the Vela pulsar[17] due to the associated remnant and/or Gum Nebula.

Finally, electron column-density fluctuations such as those suggested for the DIM have been observed directly, via optical and UV absorption-line studies. HD27127, a bright double star behind the Vela remnant, shows several high velocity components in Ca H and K absorption. Over a period not greater than 4 years and possibly of order 2 months, a component at 15km/s was seen to appear in the absorption profile[18]. The scale implied for the absorbing parcel is $\sim 10^{14}$cm, while line strength measurements show that the gas has passed through a high velocity shock and grains have largely evaporated. Using a solar abundance the observed Ca line strength gives a column density $\sim 3 \times 10^{17}$cm^{-2}. Copernicus UV observations of CII and NII fine structure[19] indicate similar column densities and EM $\gtrsim 10$ for several nearby lines of sight.

Clearly our understanding of these dramatic refraction events is restricted by the present modest sample; current efforts to extend the radio observations will provide a relatively efficient means of focusing attention on specific lines of sight. The considerations described above suggest that only highly compact extragalactic sources are suitable background objects, and that objects behind known large scale features are promising targets. Further, sources showing past behaviour might well recur (as suggested by the case of 1502+106), although the timescale may be $\gtrsim$ several years. Radio observations of refractive scintillation and lensing should continue to play an important role, with the immersion peak of a caustic event serving as an alert to initiate programs of VLBI and rotation measure observations.

However, our picture of physical conditions in these structure will benefit most from associated optical and UV studies. The diffuse background in the vicinity of known event should show evidence of the ionized gas; coherent velocity structure in emission near old remnants may also be visible. A program of optical monitoring of absorption line profiles behind these regions would allow estimates of the size and filling factor of the refractors. In several cases, radio light curve fluctuations have been seen for relatively bright ($m_v \sim$16-17) background sources, and in these cases observation of the absorption line profile during and following the event would be of particular interest, since the gas sampled on the optical path is the same as that causing the radio refraction (up to the $\sim$ AU shift due to the mean refraction angle). Although these BL Lacs, etc. provide suitable featureless backgrounds for such studies, detection of absorption variations in the Ca column (with 10 km s^{-1} resolution) will be difficult at these magnitudes so that radio monitoring of the brightest sources may be most profitable. Additionally, the large areas of the loops and middle-aged remnants should provide a number of bright stars suitable for UV absorption line studies. Unfortunately, the modest dispersion ($\sim 10^4$) available with IUE is not really sufficient for gas at the $\lesssim 100$ km/s velocities appropriate to these objects and the resolution of the space telescope may be necessary.

If the proposed identification of the refractors with galactic shocks and loops is established, these observations will provide an interesting means of studying ionized gas in modest velocity shocks at $\sim$ AU scales. Observation of post-shock structure should be an important probe of shock dynamics and cooling, and study of the physical conditions in these dense, ionized structures can provide insight into the energization and pressurization of the local ISM, as well.

It is a pleasure to acknowledge several instructive discussions with Chris McKee and Roger Blandford. Carl Heiles and Joe Silk provided useful advice on the observations and computing support was provided by NSF grant AST 86-15816.

REFERENCES

1. Fiedler, R.L., Dennison, B., Johnston, K.J., and Hewish, A., *Nature*, **326**, 675 (1987).
2. Romani, R.W., Blandford, R.D., and Cordes, J.M., *Nature*, **328**, 324 (1987).
3. Roberts, J.A., CSIRO RPP preprint 3153 (1987).
4. Romani, R.W., Narayan, R., and Blandford, R.D., *Mon. Not. R. ast. Soc.*, **220**, 19 (1986).
5. Lyne, A.G., Manchester, D.M. and Taylor, J.H., *Mon. Not. R. ast. Soc.*, **213**, 613 (1985).
6. Reynolds, R.J., *Ap.J.*, **282**, 191 (1984).
7. Ellis, G. R. A., *Austr. J. Phys.*, **35**, 91 (1982).
8. Berkhuijsen, E.M., Haslam, C.G.T. and Salter, C.J. *Astr. Ap.*, **14**, 252 (1971).
9. Colomb, F.R. Poeppel, W.G.L., and Heiles, C. *Astr. Ap. Suppl.*, **40**, 47 (1980).
10. Shapirovskaya, N.Ya., *Soviet Astron.*, **22**, 544 (1978).
11. Hjellming, R.M. and Narayan, R., *Ap.J.*, **310**, 768 (1986).
12. Heeschen, D.S., Krichbaum, T., Schalinski, C. and Witzel, A., *A.J.*, **94**, 1493 (1987).
13. McCray, R. in Physical Processes in Interstellar Clouds, Morfill, G. and Scholer, eds. (Reidel:Dordrecht), 95 (1987).
14. McKee, C.F. and Hollenbach, D.J. *Ann.Rev.Ast.Ast.*, **18**, 219 (1980).
15. McCray, R., Stein, R.F., and Kafatos, M., *Ap.J.*, **196**, 565 (1975).
16. Cordes J.M. and Wolszczan, A., *Ap.J.*, **307**, L27 (1986).
17. Hamilton, P.A., McCulloch, P.M., Manchester, R.N., Ables, J.G., and Komesaroff, M.M., *Nature*, **265**, 224 (1977).
18. Hobbs, L.M., Wallerstein, G., and Hu, E.M., *Ap.J.*, **252**, L17 (1982).
19. Jenkins, E.B. Silk, J., and Wallerstein, G., *Ap.J. Suppl.*, **32**, 681 (1976).

REFRACTIVE SCINTILLATION OF EXTRAGALACTIC RADIO SOURCES

Wm. A. Coles
ECE, University of California, San Diego, CA 92093.

ABSTRACT

Compact extragalactic sources can display intensity fluctuations which are caused by fluctuations in the electron density of the interstellar medium. The 1/e time scale of these "refractive scintillations" can be many years. The intensity structure function $D_i(\tau) = <[I(t) - I(t+\tau)]^2>$ has been computed for various distributions of the density fluctuations in the interstellar medium. The results show that the intensity fluctuations observed for periods much smaller than the 1/e time scale are dominated by the region near the earth. Thus most observations actually sample only the local region.

I. INTRODUCTION

In the past few years our understanding of the scattering process in the interstellar medium has improved considerably. In particular it is now known that scattering can produce intensity variations on both diffractive (small) and refractive (large) scales (Rickett, Coles and Bourgois 1984). Convincing evidence exists that slow variations in the apparent intensity of many sources are due to refractive scintillation (Rickett 1986). The list includes: pulsars, meter wavelength variables, flickering sources, OH and H_2O masers, and galactic plane variables. However these sources may also show intrinsic flux variations which would be of considerable theoretical interest if substantiated.

The theory of refractive scintillation in the interstellar medium is far from complete. There are serious difficulties of mathematical, physical, astronomical and experimental nature. The widely used solutions are only the first correction to a strong scattering limit. The correction series is known to be convergent but it converges very slowly (Codona, Creamer, Flatte, Frehlich and Henyey 1986). The process which produces the density fluctuations is unknown and the form of the spatial spectrum of electron density is still in question. Observations appear to be consistent with the type of spectrum characteristic of turbulence (Rickett et al. 1984) or with a spectrum which has a steeper exponent and is non-turbulent in character (Goodman and Narayan 1985). The interstellar medium is known to be grossly inhomogeneous; the scatter in measurements of the diffractive scintillation of pulsars at the same distance is a factor of 10^4 (Cordes, Weisberg and Boriakoff 1985). The spatial scale s_r for refractive scintillation can be so large that the corresponding time scale $\tau_r = s_r / V$ is many years. Thus the observational interval T_o is often less than the time scale.

The purpose of this paper is to examine the effects of the inhomogeneous distribution of density fluctuations, recognizing that other theoretical limitations remain. The first order mathematical expression has been used and a "turbulent" spectral form has been assumed. As $T_o < \tau_r$ is often the case the intensity covariance function is not the appropriate statistic.

$$C_i(\tau) = \frac{1}{|T_o - \tau|} \int_0^{T_o - \tau} dt \, [I(t) - \bar{I}][I(t+\tau) - \bar{I}] \quad \text{where} \quad \bar{I} = \frac{1}{T_o} \int_0^{T_o} dt \, I(t) \tag{1}$$

The problem is that $\bar{I}$ is not well defined by an observation interval of less than τ_r so $C_i(\tau)$ is biassed. A more useful statistic for experimental purposes is the intensity structure function $D_i(\tau)$ which is unbiased (Rickett et al. 1984, Simonetti, Cordes and Heeschen 1985).

$$D_i(\tau) = \frac{1}{|T_o - \tau|} \int_0^{T_o - \tau} dt \, [I(t) - I(t+\tau)]^2 \tag{2}$$

© 1988 American Institute of Physics

Theoretically the expectations of $C_i(\tau)$ and $D_i(\tau)$ are simply related if $T_o \to \infty$.

$$\langle D_i(\tau) \rangle = 2[\langle C_i(0) \rangle - \langle C_i(\tau) \rangle] \tag{3}$$

The effect of the intrinsic angular size of the source θ_s is important if it exceeds the scattering angle θ_o. If the scattering is confined to a thin "screen" a distance L from the observer then $C_i(\tau)/C_i(0)$ is just the autocorrelation of the source brightness distribution $B(\theta)$ evaluated at $\theta = V\tau/L$. However if the scattering medium is uniformly distributed then $C_i(\tau)/C_i(0)$ is independent of $B(\theta)$ for both spectral models (Blandford, Narayan and Romani 1986, Coles et al. 1986).

Calculations are presented here for the intermediate case of a thick slab, for both large and small sources. The results show that the level of $D_i(\tau)$ for $\tau \ll \tau_r$ is very sensitive to the level of "turbulence" near the observer. This is true both for large and small sources. Consequently an observation confined to a period $T_o \ll \tau_r$ can only probe the "local" interstellar medium.

II. THEORY

The theory to be used here is a first-order strong-scattering plane-wave approximation and applies to isotropic power-law density spectra of the form

$$\Phi_N(\vec{q}) = C_N^2 q^{-\alpha-2} \exp[-(ql_o/2)^2] \quad \text{where} \quad 1 < \alpha < 2 \ . \tag{4}$$

With this definition the Kolmogorov spectral exponent is $\alpha = 5/3$. The cutoff parameter l_o is called the "inner scale". The calculations could easily be extended to deal with anisotropic spectra. When the scattering is confined to a small forward-angle the electric field coherence $\Gamma(s) = \langle E(x)E^*(x+s) \rangle$ can be written exactly as $\Gamma(s) = \exp[-\tfrac{1}{2}D(s)]$ where

$$D'(\vec{s}, z) = 4\pi r_e^2 \lambda^2 \int_{-\infty}^{\infty} d\vec{q}\, \Phi_N(\vec{q}, z)[1 - \cos(\vec{q}\cdot\vec{s})] \tag{5}$$

$$D(\vec{s}) = \int_0^L dz\, D'(\vec{s}, z) \tag{6}$$

Here r_e is the classical electron radius and λ is the wavelength. Thus one can define an electric field coherence scale (half width at $\sqrt(e)$) s_o by $D(s_o) = 1$. The angular spectrum of plane waves is the Fourier transform of $\Gamma(\vec{s})$ so the half width at $\sqrt(e)$ of the angular spectrum is $\theta_o = 1/ks_o$ where $k = 2\pi/\lambda$.

The covariance of intensity, to first order, consists of three terms, two diffractive and one refractive in origin. As the scale of the diffractive terms is very small no sources other than pulsars have ever been observed to show diffractive scintillations. Thus the refractive term is the most important here. It is given by Codona and Frehlich (1986) for an incident plane wave.

$$C_{rp}(\vec{s}) = 8\pi r_e^2 \lambda^2 \int_0^L \int_{-\infty}^{\infty} d\vec{q}\, dz\, |V(\vec{q}(L-z)/k)|^2 \Phi_N(\vec{q}, z) \sin^2(q^2(L-z)/2k) \tag{7}$$

$$\times \exp[-\int_0^L D'(\vec{q}g(t,z)/k, z)dt + i\vec{q}\cdot\vec{s}]$$

$$g(t,z) = L-z \quad \text{for} \quad t < z, \quad \text{and} \quad g(t,z) = L-t \quad \text{for} \quad t > z$$

The source visibility function $V(\vec{s})$ is the Fourier transform of the brightness distribution $B(\vec{\theta})$. Here we will use a simple gaussian model $B(\vec{\theta}) = \exp[-\tfrac{1}{2}(\theta/\theta_s)^2]$.

The effect of the inner scale on refractive scintillations has been discussed by Coles et al. (1986). If the inner scale $l_o < s_o$ then it has very little effect on the scintillations. However as the inner scale increases to $l_o > s_o$ it causes an increase in the intensity fluctuations which peaks at $l_o \approx \theta_o L$. If $l_o > \theta_o L$ there is very little scintillation diffractive or refractive. Analysis of pulsar observations suggests that in the interstellar medium, at least for distances greater than 100pc, $s_o < l_o < \theta_o L$ and this is the case to be considered here. Then D(s) is well approximated by

$$D(\vec{s}) = (4\pi^2/2^\alpha)\Gamma(1-\alpha/2)C_{No}^2 L r_e^2 \lambda^2 l_o^{\alpha-2} s^2 \quad \text{where} \quad C_{No}^2 = \frac{1}{L}\int_0^L C_N^2(z)dz \quad (8)$$

It is convenient to write $C_r(s)$ in terms of θ_o, to normalize the z-axis to x=z/L and to define $\gamma = l_o/\theta_o L$. Then the problem is described by three spatial scales, l_o, $\theta_o L$ and $\theta_s L$ as shown below.

$$\theta_o = r_e \lambda^2 [\Gamma(1-\alpha/2) 2^{-\alpha} l_o^{\alpha-2} C_{No}^2 L]^{\frac{1}{2}} \quad (9)$$

$$C_r(s) = 2^{\alpha-1}(1-\alpha/2)\gamma^{2-\alpha} \int_0^1 dx \frac{C_N^2(z)}{C_{No}^2} \frac{(1-x)^2}{F(x)^{2-\alpha/2}} {}_1F_1[2-\alpha/2, 1, -(s/2\theta_o L)^2/F(x)] \quad (10)$$

$$F(x) = (1-x)^2 \int_0^x \frac{C_N^2(t)}{C_{No}^2} dt + \int_x^1 \frac{C_N^2(t)}{C_{No}^2}(1-t)^2 dt + \frac{\gamma^2}{4} + (\frac{\theta_s}{\theta_o})^2(1-x)^2 \quad (11)$$

The distribution of the scattering medium is in the simple slab geometry shown in Figure 1. Here the wave originating at x=−∞ propagates through free space until it encounters a slab of thickness t* at x=0. The observer is at x=1.

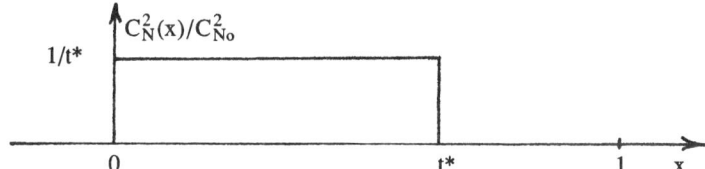

Figure 1. The distribution of turbulence studied. Here x=z/L is the normalized distance. The distance from the observer to the center of the slab is $L_{eff} = (1-t^*/2)L$.

In this case F(x) can be integrated analytically giving

$$C_r(s) = 2^{\alpha-1}(1-\alpha/2)\gamma^{2-\alpha} \int_0^1 dx \frac{(1-x)^2 {}_1F_1[2-\alpha/2, 1, (s/2\theta_o L)^2/F(x)]}{t^* F(x)} \quad (12)$$

where $F(x) = [(1-x)^2(1+2x) - (1-t^*)^3]/3t^* + \gamma^2/4 + (\theta_s/\theta_o)^2(1-x)^2$.

This is easily integrated numerically as the hypergeometric function $_1F_1$ behaves very much like a simple exponential in this parameter regime. Care must be taken near x=1 where there is a fractional order singularity.

III. CALCULATIONS

Scattering of extragalactic sources in the interstellar medium, at least at frequencies below a few GHz, is characterized by $\gamma \ll 1$ (Coles et al. 1986). Thus the term F(x) is independent of γ and $C_r(s)$ for $s > l_o$ depends on γ only through the factor $\gamma^{2-\alpha}$. In order to facilitate comparison between different screens the results have been scaled to $L_{eff} = L(1-t^*/2)$ and $\gamma_{eff} = l_o/\theta_o L_{eff}$ has been kept constant. For the calculations shown here

$\gamma_{eff} = 0.001$. The results can be scaled to a different inner scale l_o or scattering angle θ_o using the factor $\gamma^{2-\alpha}$, provided that γ remains $\ll 1$. The spectral exponent $\alpha = 5/3$ corresponding to a Kolmogorov turbulence spectrum has been used throughout. Of the class of "turbulent" spectral models this appears to give the best fit to pulsar data (Coles et al. 1986). Calculations have been done for both large and small sources and can easily be scaled to any source diameter.

A numerical solution of $C_r(s)$ for a small source ($\theta_s \ll \theta_o$) is plotted in Figure 2. One can see that $C_r(s > \theta_o L_{eff})$ is virtually unaffected by the screen thickness, but that $C_r(s \to 0)$ is quite strongly dependent on the region near the observer. This is shown more clearly in Figure 3 where $D_i(s)$ has been plotted versus $s/\theta_o L_{eff}$ on log-log axes. The 1/e width s_r of $C_i(s)$ has been marked on each plot for reference. Note that s_r varies from 1.8 $\theta_o L_{eff}$ to 1.5$\theta_o L_{eff}$ as the slab changes from "thin" to "thick". The slab t*=0.1 is indistinguishable from t*=0 on this plot. Clearly the 10% of the medium closest to the observer contributes almost all of the intensity variation for $\tau < \tau_r$. If the plot had been continued to still smaller spacings a break to $D_i(s) \approx s^2$ would have been seen at $s = l_o$.

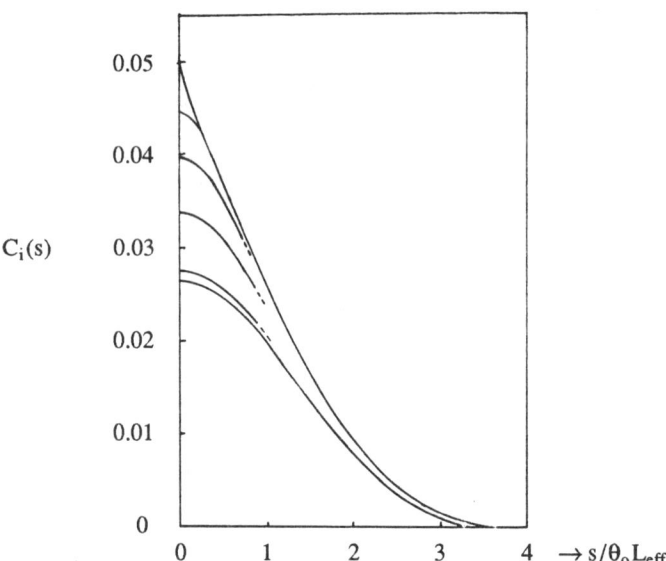

Figure 2. The refractive component of the intensity covariance. Here the source diameter is negligable, the spectral exponent is $\alpha = 5/3$ and $(l_o/\theta_o L_{eff}) = 0.001$. The thickness of the slab t* = 1.0, 0.9, 0.75, 0.5, 0.1 and 0 from top to bottom.

In many cases, perhaps most, the source diameter of extragalactic sources will not be negligable. A similar calculation for $\theta_s = 3\theta_o$ is displayed in Figure 4. The result is surprisingly similar to that of Figure 3. Again the 10% of the medium closest to the observer dominates the variation for $\tau < \tau_r$ (or $s < s_r$). The 1/e widths s_r run from 1.9$\theta_s L_{eff}$ to 1.5$\theta_s L_{eff}$. In general a good approximation for the scale is

$$s_r = 1.6\sqrt{\theta_o^2 + \theta_s^2}\, L_{eff} \qquad (13)$$

The main effect of increasing the source diameter is that the variance is reduced by the factor $(\theta_o/\theta_s)^{\alpha-4}$ and the scale is increased by the factor θ_s/θ_o. This calculation can easily be scaled to other values of θ_s using these factors.

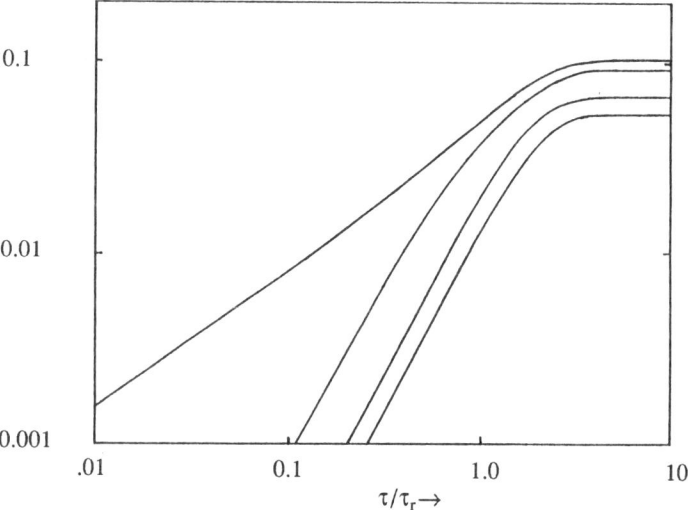

Figure 3. The structure function of intensity $D_i(s) = 2[C_r(0) - C_r(s)]$ for the "point" source calculation displayed in Figure 2. The thickness of the slab $t^* = 1.0, 0.9, 0.5$ and 0 from top to bottom.

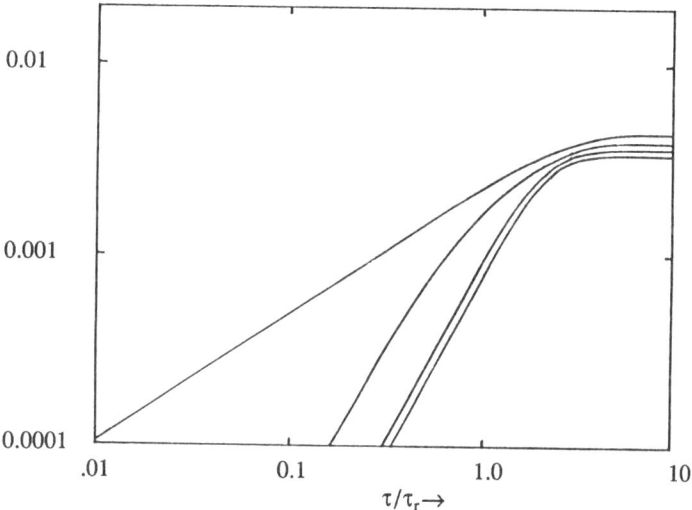

Figure 4. The structure function of intensity $D_i(s)$ for the same conditions as in Figure 3 but with a large source $\theta_s = 3\theta_o$.

IV. DISCUSSION

These calculations can be compared with, for example, the flickering observations of Simonetti et al. (1986). They have observed that intensity structure functions of flat spectrum sources tend to increase linearly with τ. Such observations are compatible with interstellar scintillation if the scattering medium is uniformly distributed, particularly near the earth. However they are incompatible with a model in which the scattering is caused by a few localized regions of intense turbulence, unless the earth is in such a region.

Another interesting comparison is with the suggestion of Rao (private communication 1986) that compact extragalactic sources near the galactic plane should show variability in their 327MHz fluxes over a time interval of 10 years. At this frequency one might expect a scattering angle of 1" near the galactic plane. Using L=10kpc and V=50km/s yields a time scale of $\tau_r \approx 200\sqrt{1+\theta_s^2}$ where θ_s is in arc sec and τ_r is in years. Thus 10 years is, at best, very much less than τ_r. If the source is less than 1" the expected rms flux difference over 10 years is just the square root of $D_i(\tau/\tau_r = 0.05)$. This would be 4% if the medium is uniformly distributed but only 0.5% if the medium is deficient near the earth.

Observations (Rao and Ananthakrishnan 1984) suggest that there is enhanced scattering in the direction of the galactic center and this certainly is a reasonable hypothesis. Thus in this direction a "thin" model would be appropriate and one would not expect to see any time variation at 327MHz over accessible time scales.

Acknowledgements

This work was done while the author was visiting the Radio Astronomy Centre of the Tata Institute of Fundamental Research at Ootacamund, India. I would particularly like to acknowledge the help of A. Pramesh Rao and the staff at Ooty. The visit was supported by the NSF International Programs Office under NSF-INT-8520635.

References

Blandford R., Narayan, R. and Romani, R. W. 1986, Ap. J. (Letters), 301, L53.
Codona, J. L., Creamer, D. B., Flatte, S. M., Frehlich, R. G. and Henyey, F. S. 1986, Radio Sci. 21, 929.
Codona, J. L. and Frehlich, R. G. 1987, Radio Sci. 22, 469.
Coles, W. A., Rickett, B. J., Frehlich, R. G. and Codona, J. L. 1987, Ap. J., 315, 666.
Cordes, J. M., Weisberg, J. M. and Boriakoff, V. 1985, Ap. J. 288, 221.
Goodman, J. and Narayan, R. 1985, M. N. R. A. S., 214, 519.
Rao, A. P. and Ananthakrishnan, S. 1984, Nature, 312, 707.
Rickett, B. J., Coles, W. A. and Bourgois, G. 1984, Astr. Ap., 134, 390.
Rickett, B. J. 1986, Ap. J., 307, 564.
Simonetti, J. H., Cordes, J. M. and Heeschen, D. S. 1985, Ap. J., 296, 46.

EFFECTS OF TURBULENT INTERSTELLAR CLOUDS ON REFRACTIVE SCINTILLATION

R.G. Frehlich

Cooperative Institute for Research in the Environmental Sciences (CIRES)
University of Colorado/NOAA, Boulder, CO 80309

ABSTRACT

The interstellar medium is believed to consist of turbulent clouds embedded in a more uniform turbulent medium. Some of the effects of this model will be determined by theoretical calculations of the refractive covariance using a Kolmogorov turbulence spectrum with an inner scale. Three physical regimes will be discussed.

INTRODUCTION

Recently, the large scale component of scintillation was identified[1] as the process responsible for the slow fluctuations of pulsar flux[2-4]. The justification for the refractive component of strong scintillation was based on calculations[5] of the intensity covariance for a Kolmogorov turbulence spectrum which predicted the correct scales but insufficient amplitude[1]. An enhancement in refractive amplitude can be justified by many different models for the turbulence spectrum[6-10].

The nature of interstellar scintillation[11,12] suggests that the interstellar medium is composed of clumps of turbulent clouds superimposed on a background of turbulence that depends on the Galactic location. A more complete understanding of the space-time statistics of scintillation is required in order to unravel the complicated turbulence structure of the interstellar medium and its' effects on radio astronomical measurements and interpretations.

REFRACTIVE SCINTILLATION

A heuristic analysis[6] and a geometrical optics analysis[13] has provided a basic understanding of the scintillation process. A more rigorous expression for the refractive component of the space-time intensity covariance is given by a series solution[14,15] where the leading term is

$$C_{rs}(\vec{\beta},\tau) = 8\pi r_e^2 \lambda^2 \int_0^L dz_1 \int_{-\infty}^{\infty} d\vec{q} |V[\frac{\vec{q}}{k}(L-z_1)]|^2 \sin^2[\frac{q^2}{2k} h(z_1,z_1)]$$

$$\Phi_{N_e}(\vec{q},z_1) \exp\left[-\int_0^L D'[\frac{\vec{q}}{k} h(z,z_1),z] dz\right] \exp\{i\vec{q}\cdot[\vec{\beta}\frac{z_1}{L}-\vec{V}(z_1)\tau]\} \quad (1)$$

Here, λ is the wavelength of the radiation, k is the wave number, r_e is the classical electron radius, $\vec{\beta}=\vec{x}_1-\vec{x}_2$ is the transverse separation of two observation points at a distance L from the extended source, $V(\vec{\beta})$ is the normalized visibility function, $\tau = t_1 - t_2$ is the separation in time, $\vec{V}(z)$ is the transverse velocity distribution as a function of the propagation coordinate z, $\Phi_{N_e}(\vec{q}, z)$ is the electron density spectrum at position z, and

$$h(z,z_1) = z(\frac{z_1}{L} - 1) \quad z < z_1$$

$$= z_1(\frac{z}{L} - 1) \quad z > z_1 \qquad (2)$$

The case of a distant source (extra galactic) viewed through an extended random medium is obtained as a limit of this expression. The relative amplitude of the refractive fluctuations m_r is given by

$$m_r^2 = C_r(\vec{\beta} = 0, \tau = 0) \qquad (3)$$

The Born approximation is obtained if the first exponential term of (1) is negligible. This exponential term describes the decorrelation due to scattering from irregularities larger than the Fresnel surface. This is the physical explanation of the exponential cut-off in wave number $\vec{q}$ and provides a description of the effective scattering region.

We consider the point source case (pulsar scintillation) and the turbulence spectrum[10,16,17]

$$\Phi_{N_e}(q) = C_N^2 q^{-\alpha - 2} \exp[-(ql_0/2)^2] \qquad (4)$$

where l_0 is the inner scale and C_N^2 is the "level of turbulence" (the Kolmogorov slope is $\alpha = 5/3$). When the inner scale is larger than the coherence length of the received field s_0 the angle of the scattered field is

$$\theta_0 = \frac{1}{ks_0} = [\Gamma(1 - \alpha/2) C_N^2 2^{-\alpha} L/3]^{1/2} r_e l_0^{\alpha/2 - 1} \lambda^2 \qquad (5)$$

If the turbulence is uniform, the exponential term of (1) becomes

$$\exp\left[-\int_0^L D'[\frac{\vec{q}}{k} h(z, z_1), z] dz\right] = \exp\left[-[x(1-x)\theta_0 L q]^2\right] \qquad (6)$$

where $x = z/L$ is the normalized propagation distance. This term defines the width of the effective scattering region $S(x)$ as

$$S(x) = x(1-x)\theta_0 L \qquad (7)$$

The width of the effective scattering region for the case of an extra galactic source (plane wave geometry) and a uniformly turbulent galaxy is obtained from the plane wave limit of (6), i.e.

$$S(x) = [(1 + 2x)/3]^{1/2}(1-x)\theta_0 L \qquad (8)$$

Both cases are plotted in Fig. 1.

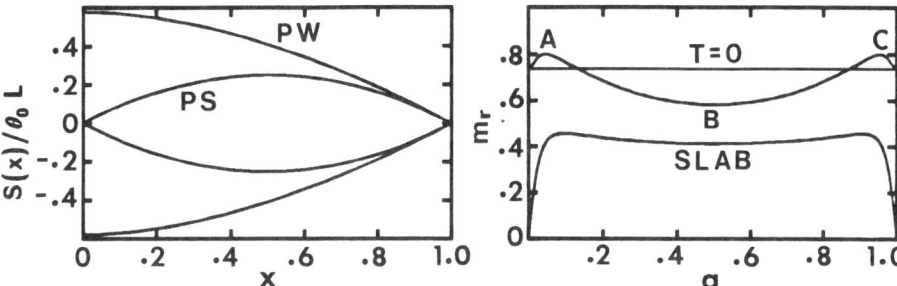

Fig. 1. The effective scattering region for the point source (PS) and plane wave (PW) geometry.

Fig. 2. The refractive modulation m_r as a function of slab location a for the profile (9), the SLAB only, and no slab (T=0).

The behavior of the scattering region as a function of the turbulence profile is essential for a complete understanding of the scintillation process.

EFFECTS OF A TURBULENT CLOUD

Measurements of pulsar scintillation indicate that the interstellar medium is composed of turbulent clouds superimposed on a turbulent background[11,12]. All the analysis presented here assumes the interstellar medium has homogeneous statistics in the direction transverse to the propagation over dimensions comparable to the width of the effective scattering region $S(x)$.

A turbulent medium with one added turbulent cloud can be modeled by the C_N^2 profile

$$C_N^2(z) = C_N^2 \qquad z<(a-b/2)L, \; z>(a+b/2)L$$
$$= C_N^2(\frac{T}{bL}+1) \quad (a-b/2)L<z<(a+b/2)L \qquad (9)$$

where aL is the location of the slab and bL is the slab thickness. The refractive covariance (1) for strong scattering and $l_0>s_0$ reduces to a one-dimensional integral[10]. The rms refractive amplitude is shown in Fig. 2 as a function of slab location for the profile (9), SLAB only, and no slab (T=0). Here, the frequency is 1.2 GHz, L=500 pc, $C_N^2=.001 m^{-20/3}$, $l_0=10^9$ meters, b=.001, and T=1. The maxima and minima of Fig. 2 identify three mechanisms (A,B,C) that govern refractive scintillation. The physics of these three mechanisms is related to the width of the effective scattering region $S(x)$.

Region A- The maximum enhancement of refractive variance occurs when the slab location is such that $S(x)=l_0$, i.e., one inner scale is illuminated in the scattering region (Fig. 3, note that the scattering region is only slightly altered by the addition of the slab). Thus, one lens focuses intensity into the large scale structure. The covariance function for this case is plotted in Fig. 4. The covariance at small spacing is not affected by the addition of the slab but their is a marked enhancement at large separation.

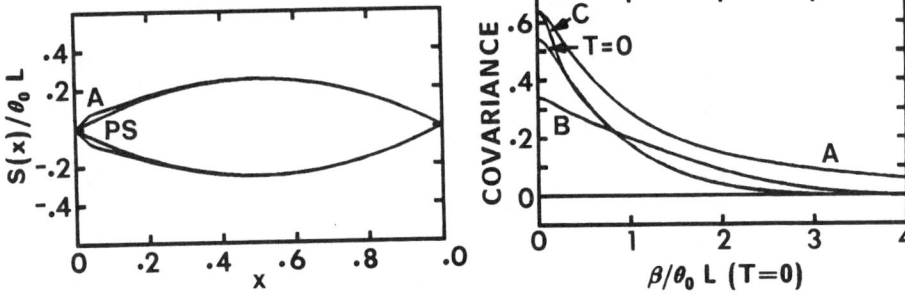

Fig. 3. The effective scattering region for mechanism A and the point source (PS) with no turbulent slab.

Fig. 4. The covariance functions for the case of uniform turbulence (T=0) and mechanism A, B, and C.

Region B- The minimum in refractive variance is due to the increase in the scattering region (Fig. 5). The irregularities with dimensions of the scattering region produce refractive structure of the same scale. The larger the number of individual scattering sites of dimension l_0, the smaller the resulting refractive variance. This is the interpretation presented in Ref. 1. The covariance function for this case is plotted in Fig. 4.

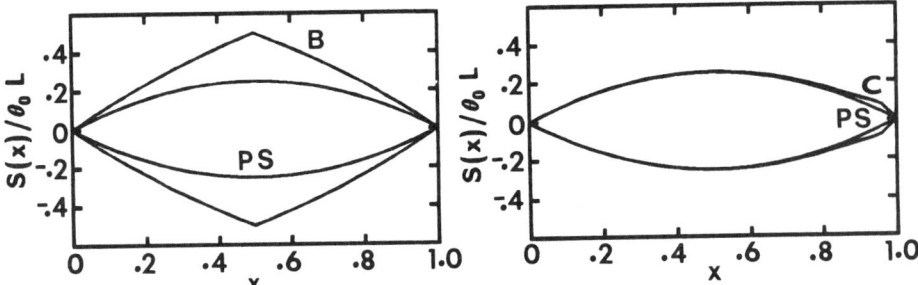

Fig. 5. The effective scattering region for mechanism B and the point source (PS) with no turbulent slab.

Fig. 6. The effective scattering region for mechanism C and the point source (PS) with no turbulent slab.

Region C- The second maximum of refractive variance again occurs where l_0 has the same dimensions (Fig. 6) as the width of the effective scattering regon $S(x)$. The focusing of energy by these lenses produces a small scale structure of dimensions l_0. This structure also has large bandwidth and is not affected by sources that have angular dimensions less than the scattering angle θ_0. The temporal scale is the time it takes the irregularities of size l_0 to move out of the scattering region $\approx l_0/V$. This mechanism has intriguing implications for line-of-sights that pass through a turbulent cloud near earth.

Refractive scintillations of much smaller spatial and temporal scales would be produced. The covariance function for this case is plotted in Fig. 4.

The planar geometry (extra galactic sources) will not have Mechanism A since the scattering region is not small at the beginning of the random media (see Fig. 1). This explains why the refractive fluctuations are larger for a spherical geometry than a plane wave geometry. The spherical geometry contains an extra focusing mechanism.

The effects of source size and the velocity of the clouds on the space-time intensity covariance is a simple addition to the theory. The effects of two or three clouds can also be investigated.

REFERENCES

1 B. J. Rickett, W.A. Coles, and G. Bourgois, Astron. Astrophys. **134**, 390-395 (1984).

2 D. J. Helfand, L.A. Fowler, and J.V. Kuhlman, Astron. J., **82**, 701-705 (1977).

3 W. Sieber, Astron. Astrophys., **113**, 311-313 (1982).

4 O. B. Slee, S.K. Alurkar, and A.D. Bobra, Aust. J. Phys., **39**, 103-114 (1986).

5 A. M. Prokhorov, F.V.Bunkin, K.S. Gochelashvily, and V.I.Shishov, Proc. IEEE,**63**, 790-811 (1975).

6 R. Blandford and R. Narayan, Mon. Not. R. Astro. Soc., **213**, 591-611 (1985).

7 J. Goodman and R. Narayan, Mon. Not. R. Astro. Soc., **214**, 519-537 (1985).

8 R. W. Romani, R. Narayan, and R. Blandford, Mon. Not. R. Astro. Soc., **220**, 19-49 (1986)

9 N. Y. Shapirovskaya and W. Sieber, Astron. Astrophys., **136**, 171-174 (1984).

10 W. A. Coles, R.G. Frehlich, B.J. Rickett, and J.L. Codona, Astrophys. J., **315**, 666-674 (1987).

11 B. J. Rickett, Ann. Rev. Astron. Astrophys. **15**, 479-504 (1977).

12 J. M. Cordes, J.M. Weisberg, and V. Boriakoff, Astrophys. J., **288**, 221-247 (1985).

13 B. J. Rickett, Astrophys. J., **307**, 564-574 (1986).

14 R. G. Frehlich, Radio Sci., **22**, 481-490 (1987).

15 J. L. Codona and R.G. Frehlich, Radio Sci., **22**, 469-480 (1987).

16 J. C. Higdon, Astrophys. J., **285**, 109-123 (1984).

17 J. C. Higdon, Astrophys. J., **342**, 342 (1986).

REFRACTION FROM INTERSTELLAR SHOCKS

ANDREW W. CLEGG[*], DAVID F. CHERNOFF[†]
AND JAMES M. CORDES[*†]

[*]National Astronomy and Ionosphere Center
[†]Center for Radiophysics and Space Research
Cornell University, Ithaca, N.Y. 14853

ABSTRACT

Interstellar shocks that are sufficiently thin ($\lesssim 10$ au) and sufficiently strong ($\Delta DM \gtrsim 0.01$ cm^{-3}pc) are capable of producing observable flux modulations and multiple images of background point radio sources. From the statistics of observed refraction events, inferences may be drawn concerning the relationship between small-scale electron density structure and large-scale phenomena like supernova explosions and strong stellar winds. In order for shock refraction events to be frequent, the effective cross section must be much larger than predicted for an ideal, spherical shock. The cross section is likely to be enhanced by turbulence that distorts the shock interface on scales of ~ 1 au and by the prevalence of HI clouds and other non-uniformities in the ISM that will be swept up or engulfed by the expanding blast wave.

I. INTRODUCTION

The prevalence of high energy phenomena implies that the interstellar medium is pervaded by shocks. Considering the various methods of creating shocks, including strong stellar winds and supernova explosions, the likelihood of encountering a remnant shock in any given volume of space, or along any given line of sight, may be relatively high.

Only the strongest shocks can be directly detected by Earth-based observations. In this paper we report a method of *indirect* detection, where the effect of the shock on background objects can be studied. Lensing action of the shock causes intensity variations, shifts in apparent position, multiple images, and other aberrations, whose amplitudes depend on the strength and thickness of the shock. Observations along a sufficiently large number of lines of sight may allow general conclusions to be drawn regarding the structure and abundance of shocks in the ISM. In the following we take the point of view that refraction events are seen and we estimate the conditions under which shocks may account for the structure and frequency of occurrence of the events.

II. THEORY

THE IDEAL SHOCK. For a rough approximation, an interstellar shock is modeled by an infinitely thin screen lying a distance D from the observer. The screen is divided into a region of dispersion measure $DM = 0$ and a region of $DM = DM_0$, separated by a transition region of thickness $l_\perp$. For mathematical convenience, a dispersion measure profile across the screen, perpendicular to the line of sight, is taken to be of the form $DM(x) = DM_0/(e^{-x/l_\perp} + 1)$, where the middle of the transition region (*i.e.*, the shock front) is located at $x = 0$. The geometry is shown in figure 1.

The screen advances the phase of an incident plane wave by an amount $\Delta\phi(x) = \lambda r_e DM(x)$, where λ is the wavelength of observation and r_e is the classical electron radius. The wave front emerging from the screen is no longer planar, but has a surface of constant phase that is the inverse of the dispersion measure profile. In the limit of geometrical optics, rays propagating normal to the constant phase surface delineate the flow of energy.

Figure 2 is a ray trace performed on an ideal shock for which $l_\perp = 1.0$ au and $DM_0 = 0.01$ cm^{-3}pc. The shock is located along the $y = 0$ plane, with the middle of the shock at $x = 0$. The shock produces refractive focusing and defocusing of the incident plane wave, as indicated by the convergence and divergence of rays. An observer moving with respect to the shock will notice flux variations of the background source upon passing through focusing and defocusing regions. At a distance $D_c \propto l_\perp^2/\lambda^2 DM_0$, ray crossing occurs and multiple images form. The multiple-image region is surrounded by a caustic surface, upon which the image intensity grows without bound in the limit of geometrical optics. The actual intensity is limited by diffraction, which has not been taken into account here.

THE REFINED SHOCK. We have incorporated the results of detailed shock calculations to construct a schematic model of the electron density variation. We discuss the canonical case with the shock velocity v_s of 100 km s^{-1}, the hydrogen nuclei density n_0 of 10 cm^{-3}, and magnetic field strength B of 3 μG. These typical parameters should apply to supernova shocks with explosion energies of 10^{51} ergs which have propagated for about 10^5 years into a uniform interstellar medium. Elsewhere we will consider wider ranges of initial conditions.

The electron number density is estimated in four regions. (1) Upstream of the shock, material is pre-ionized by the photon flux from the hot postshock regions. Shull and McKee (1979) tabulate the flux at the shock front, and we integrate the transfer equations upstream taking into account photoionization and recombination. The H nuclei density is fixed, while a Strömgren-like region forms ahead of the shock. (2) At the shock jump (j-shock) the density rises by a factor of 4, the hydrogen is fully ionized, and the temperature reaches $T_s = 1.38 \times 10^5$ K. The transition is smoothed on an atom-atom mean free path; the structure of the shock is unknown but may be far shorter if governed by plasma processes. (3) The cooling behind the shock for $T > 10^4$ K is estimated from a crude fit to the calculations of Shull and McKee, $\mathcal{L} \approx \mathcal{L}_0 T^a$, where $\mathcal{L}_0 \approx 10^{-23}$ erg cm^3 s^{-1} and $a \approx 2.0$. The density is solved for approximately (eqn. 2.37 of Hollenbach and McKee 1979) for $T > T_m$, where $T_m \approx 10^4$ K is the temperature at which the magnetic field pressure begins to exceed the thermal pressure and limit the compression. For our canonical case, T_m is also the temperature at which significant recombination begins. The ionization fraction is assumed fixed at the immediate postshock value for $T > T_m$. (4) For $T < 10^4$ K the ionization fraction is determined by solving the downstream radiative transfer equation, in a manner analagous to region (1). The density is assumed fixed, since the field pressure dominates thermal pressure.

The dispersion measure profiles have two prominent features: the hydrodynamic discontinuity in n_e and, farther downstream, a peak (or "cusp") in n_e (see figure 3). The peak results when gas compression ceases on account of the magnetic field, and the gas begins to recombine. Although our survey is not extensive, it appears that the pre-shock ionization is never an important element.

Ray traces were performed with the refined profile. Electron densities were converted to dispersion measures by assuming an effective line of sight pathlength through the shock of 10 au. The results may be characterized as the superposition of two lensing events similar to that seen in figure 2. Lensing due to the j-shock is less likely to be seen than that from the cusp since the amplitude of the j-shock is less. The electron density structure at the cusp causes its refraction properties to mimic those of a diverging lens.

III. THE OBSERVABILITY OF REFRACTION EVENTS

Shocks are generated in supernova explosions and strong stellar winds. The abundances of either type of shock are not completely unrelated; the massive young stars capable of producing strong winds are generally considered to be supernova progenitors.

The mean free path between remnants is $l_{mfp} = (n_s \bar{\sigma}_s)^{-1}$, where n_s is the number density of remnants and $\bar{\sigma}_s$ is the mean effective cross section of each remnant. Here we take $\bar{\sigma}_s \simeq \pi(R_{\max}/2)^2 \simeq v_s^2 \tau_s^2$, where $R_{\max}$ is the maximum shock radius, which has been expressed in terms of the average shock expansion velocity v_s and lifetime τ_s. The number density of remnants in terms of the volume supernova rate $\dot{n}_{sn}$ is approximately $\dot{n}_{sn}\tau_s$. Taking $v_s = 100$ km s^{-1}, $\tau_s = 10^6$ yr, and using Spitzer's (1978) value of $\dot{n}_{sn} = 5 \times 10^{-13}$ yr^{-1}pc^{-3}, $l_{mfp} \simeq 200$ pc. Thus for a pathlength $L = 1$ kpc, the chance of intersecting at least one shock is 99.3%, while typically $\bar{n} = L/l_{mfp} \simeq 5$ shocks will be encountered.

The frequency with which refraction events occur depends on the scale of electron density structure within the shock. Refraction is observed only when the gradient in electron density perpendicular to the line of sight is sufficiently large. For a uniform shock, the gradient is maximum when the line of sight is tangent to the shock front. The frequency with which this alignment occurs is $dN/dt \sim \bar{n}/\tau_s$. For $\bar{n} \simeq 5$ and $\tau_s \simeq 10^6$ yr, $dN/dt = 5 \times 10^{-6}$ yr^{-1}, or one refraction event per 200,000 years. If stellar wind shocks are included, dN/dt is increased by approximately a factor of two.

It is likely that embedded turbulence and nonuniformities within the shocks substantially increase the number of observable events. If electron density corrugations occur on a length scale a within the shock, for example, then $dN/dt \to f dN/dt$, where $f \simeq R_{\max}/a$. Evidently if a is a few pc to a few au, dN/dt is enhanced by a factor of $10^2 - 10^5$, bringing the rate at which we expect refraction events for a source 1 kpc distant to $\sim 10^{-4} - 1$ yr^{-1}.

If a typical refraction event has duration τ_r yr, then the likelihood that an instantaneous observation along a particular line of sight reveals an event is $\sim \tau_r dN/dt$. In our calculations, τ_r is $\sim 0.1 - 1$ yr for $v_s = 100$ km s^{-1}, thus the possibility of seeing an event is $\gtrsim 10^{-7} f$ for a given line of sight.

IV. DISCUSSION

A line of sight to a source more than 200 pc distant is likely to intersect a shock. Whether or not the shock has an observable effect depends on many factors. For example, it was noted earlier that the distance at which ray crossing occurs is $D_c \propto l_\perp^2/\lambda^2 DM_0$. If the shock is too thick or weak, D_c can become very large, and an observer is unlikely to pass through a region in which refractive modulations are significant. A general rule of thumb is that strong refraction (i.e., multiple imaging) will occur if $D(kpc)\lambda^2(cm^2)DM_0(cm^{-3}pc)/l_\perp^2(au^2) \gtrsim 4$.

Non-uniformities within a shock *increase* the effective cross section for strong refraction events. Electron density fluctuations caused by embedded knots of supernova ejecta will produce additional refraction, as will smaller-scale shocks created when interstellar HI clouds are consumed by an expanding shock front. Corrugations will also act to increase f.

Although the total expected rate for refraction events towards any given source depends on several complex factors, monitoring programs fix an approximate value for f. Fiedler et al. (1987) noted anomalous flux variations of three extragalactic sources in observations covering approximately 200 source-years in a manner that is consistent with refraction through a structure with $DM_0 \lesssim 1$ cm^{-3}pc and $l_\perp \sim 1-10$ au. Durations of the events were $\tau_r \sim 0.2$ yr, comparable to the time scales expected from our shock calculations. One of the observed lensing events shows structure in frequency and time that is clearly consistent with a diverging Gaussian lens [Romani et al. (1987)]. As noted earlier, refraction from the cusp of our theoretical shock is similar to that expected from such a lens: two regions of ray convergence or crossing, separated by a defocusing region. Romani et al. further point out that these structures are overpressured with respect to the surrounding ISM, which suggest that they are expanding shocks. The value of $dN/dt \simeq 10^{-2}$ yr^{-1} from Fiedler et al. indicates that $f \simeq 10^4$, corresponding to electron density structure within the shocks on scales of $10^3 - 10^4$ au.

This research was supported by NSF grants AST 86-57467, AST 84-15162, and AST 85-20530 to Cornell University.

REFERENCES

Fiedler, R. L., Dennison, B., Johnston, K. J., and Hewish, A. 1987, *Nature*, **326**, 675.

Hollenbach, D. J., and McKee, C. F. 1979, *Ap.J. Supp.*, **41**, 555.

Romani, R. W., Blandford, R. D., and Cordes, J. M. 1987, *Nature*, **328**, 324.

Shull, J. M., and McKee, C. F. 1979, *Ap.J.*, **227**, 131.

Spitzer, L. 1978, *Physical Processes in the Interstellar Medium* (New York: Wiley & Sons).

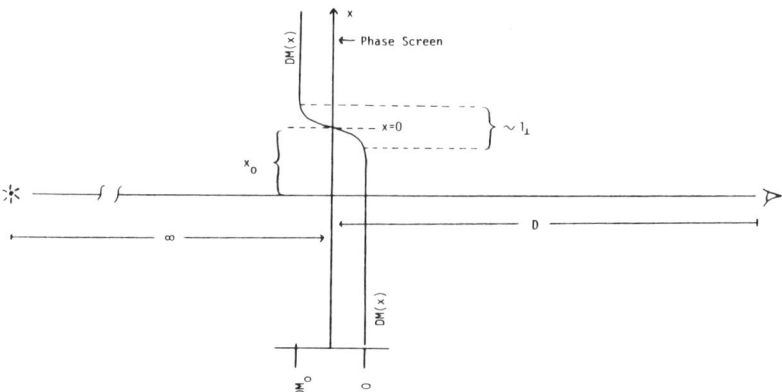

Figure 1. Viewing geometry. An infinitely distant point source illuminates the shock screen, which is located a distance D from the observer.

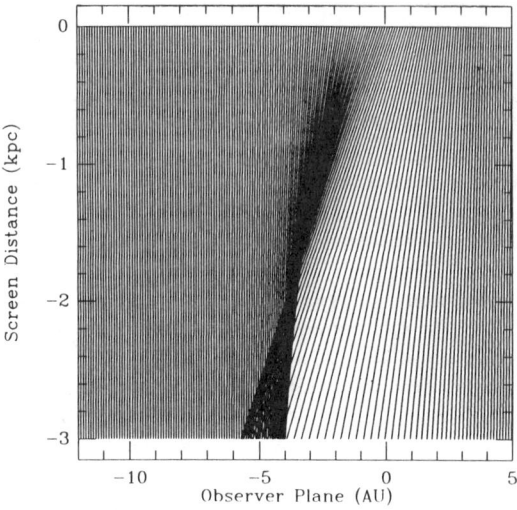

Figure 2. Ray trace for an ideal shock, with $l_\perp = 1.0$ au, $DM_0 = 0.01$ cm^{-3} pc. The y-axis indicates distance from the screen in kpc; the x-axis is distance parallel to the screen in au. A plane wave is incident from above ($y > 0$), and passes through a shock centered on the origin. A region of intense focusing is evident $\sim$ 1.4 kpc from the screen, while ray crossing, and hence multiple images, occur for distances $\gtrsim 1.7$ kpc.

Figure 3. (*Top*) Electron density profile for the refined shock. Numbers correspond to regions described in text. (*2nd from top*) Corresponding ray trace, as in figure 2. (*Bottom*) Profiles along the $y = -3$ kpc plane of, from top to bottom, total flux (linear scale), ratio of fluxes from second- and first-brightest images, and the apparent separation (in mas) between the two images. All plots in figure 3 are reproduced to the same scale along the x-axis. The flux from the background source is normalized to 1 in the absence of a refracting screen.

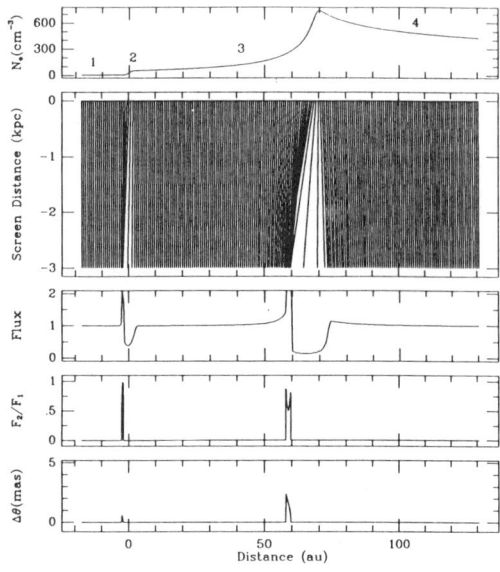

VI. THE GALACTIC DISTRIBUTION OF SCATTERING MATERIAL

GALACTIC DISTRIBUTION OF ELECTRON DENSITY TURBULENCE

J. M. Cordes[1], S. R. Spangler[2], J. M. Weisberg[3], T. R. Clifton[4]

ABSTRACT

Scintillation and angular broadening measurements are used to estimate the spectral coefficient C_n^2 of the electron density power spectrum. Relevant length scales are $\lesssim 10^{11}$ cm. Line of sight averages of C_n^2 vary by almost 5 orders of magnitude and it is likely that *local* variations are much larger. Study of 150 lines of sight supports the existence of a diffuse distribution of scattering material with large scale height ~ 0.5 kpc combined with clumps of intense turbulence having a small scale height ~ 0.1 kpc. The diffuse component appears to have a galactocentric radial extent of about 7 kpc. The clumped medium can account for the scattering of recently discovered pulsars in the inner Galaxy and of the galactic center source, Sgr A*.

INTERPRETING SCINTILLATION AND SCATTERING DATA

We assume, as is common[1], that electron density variations δn_e are described by a power law spectrum

$$P_{n_e}(q) = C_n^2 q^{-\alpha} \qquad q_0 \leq q \leq q_1, \tag{1}$$

with a spectral index α between cutoff wavenumbers $q_{0,1}$. In the following, we assume the Kolmogorov index, $\alpha = 11/3$. Observable quantities may be derived in terms of α, the coefficient C_n^2, distance to the source D, observation frequency ν and, of course, some assumed variation of C_n^2 along the line of sight. We discuss two related quantities, the scattering measure[2] (a line of sight integral of C_n^2)

$$SM \equiv \int_0^D d\ell \, C_n^2(\ell), \tag{2}$$

and the average value[3] of C_n^2, $\overline{C_n^2} \equiv SM/D$.

Angular diameters θ_{FWHM} caused by scattering may be inverted as

$$SM = \left(\frac{\theta_{FWHM}}{\theta_0}\right)^{5/3} \nu^{11/3} \tag{3}$$

where ν is in GHz and $\theta_0 = 0.13$ arc sec. Eqn (3) holds for extragalactic sources. Galactic sources embedded in the scattering medium require multiplication of the right hand side by $\alpha - 1 = 8/3$. Measurements of the scintillation bandwidth $\Delta \nu_d$ or, equivalently, the pulse broadening time $\tau_d = (2\pi \Delta \nu_d)^{-1}$, yield

$$SM = A_{11/3} \nu^{11/3} (D \Delta \nu_d)^{-5/6}. \tag{4}$$

The constant is $A_{11/3} = 8.4 \times 10^{-4}$ and in both equations (3) and (4), SM has units of kpc m$^{-20/3}$.

---

[1] National Astronomy and Ionosphere Center
[2] University of Iowa
[3] Carleton College
[4] University of California, Berkeley

Galactic Distribution

EVIDENCE FOR A TWO COMPONENT MEDIUM

In Figure 1, $\overline{C_n^2}$ is plotted against distance for 142 pulsars. Distances have been estimated from the dispersion measure DM and use of a model for the *mean* electron density[4]; in a few cases independent distance information has been used. Pulsars in different galactic latitude ranges are plotted with different symbols. Some of the points are actually upper limits. Scrutiny of Figure 1 shows that:

- $\overline{C_n^2}$ varies by almost 5 orders of magnitude.
- The statistics of $\overline{C_n^2}$ vary strongly with galactic latitude, b. Low latitude sources have $\overline{C_n^2}$ that cover the entire observed range while high latitude sources vary much less. For $b > 5°$, most of the individual values range between $-4 \lesssim \log \overline{C_n^2} \lesssim -2$, while for $b \leq 5°$, a much larger range is seen, $-4 \lesssim \log \overline{C_n^2} \lesssim 0.5$.
- The nearby Vela pulsar shows one of the largest measured $\overline{C_n^2}$, presumably due to enhanced turbulence in the Gum nebula.
- Angular broadening of the galactic center source Sgr A* (Backer, these proceedings) shows $\overline{C_n^2}$ that is comparable to the largest values seen from pulsars. As such, the scattering toward Sgr A* is completely consistent with the pulsar data, though very much enhanced with respect to nearby pulsars.

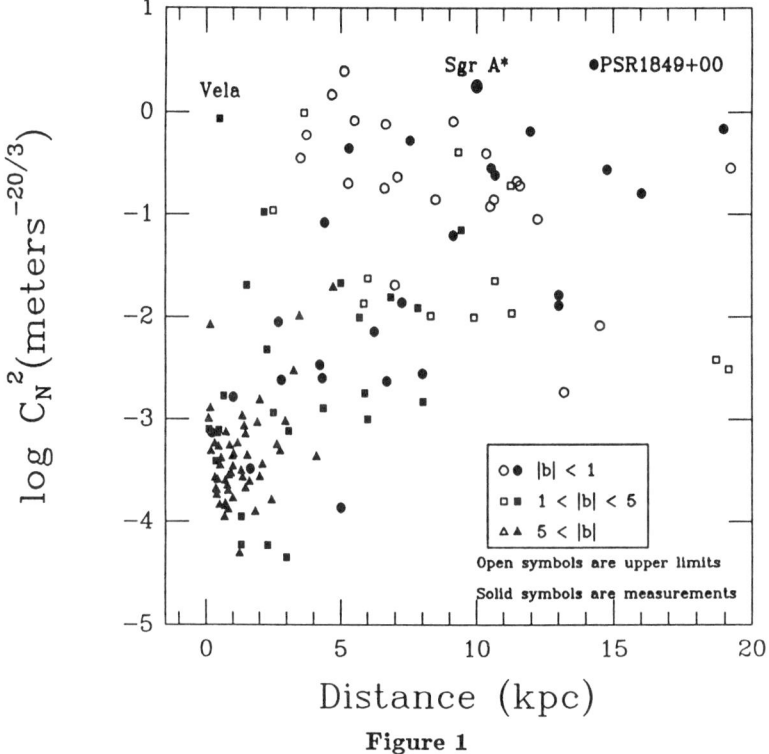

Figure 1

Figure 2 shows the latitude and longitude dependence of $\overline{C_n^2}$. The largest values of $\overline{C_n^2}$ are at low latitudes and at longitudes within 60° of the galactic center, a result also seen in ref 5.

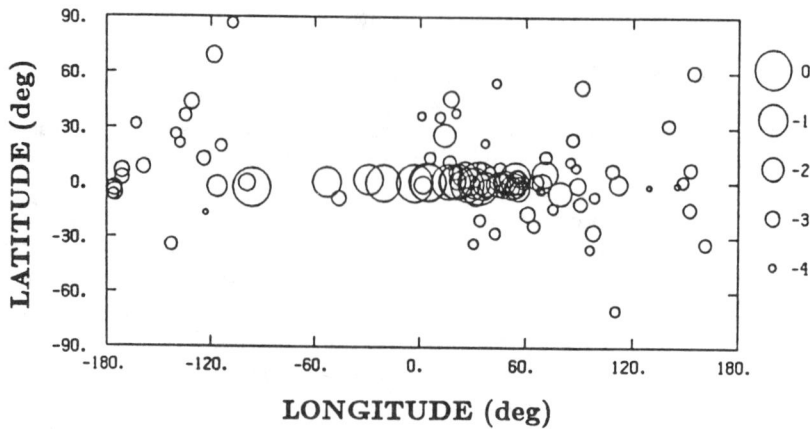

Figure 2 $\log \overline{C_n^2}$ plotted against galactic latitude and longitude for 135 pulsars. The size of the circle is linearly proportional to $\log \overline{C_n^2}$.

The variation of $\overline{C_n^2}$ with D, ℓ, and b is consistent with scattering material being distributed in a two component medium[3]. The first 'A' component has large galactic scale height and is ubiquitous while the second 'B' component has smaller scale height and is present in clumps where C_n^2 is evidently very large. Ref 3 analyzed a subset of the data in Figure 1. New observations[5,6] support the conclusions of ref 3. The analysis of ref 5 supports the 'clumpy' nature of the medium and suggests that C_n^2 increases rapidly toward the galactic center, e.g. $C_n^2 \propto R^{-3}$. Ref 5 found the distinction between the A and B components to be less well defined than ref 3. However, we feel that the statistical analysis in ref 5 may be contaminated by uncertain measurements of pulse broadening. In fact, we have designated as upper limits those cases where observations were available at only a single frequency.

CONSTRAINTS ON THE DIFFUSE, TYPE A COMPONENT

In Figure 3 we show values of the 'perpendicular' scattering measure $SM_\perp \equiv SM \sin|b|$ plotted against distance from the galactic plane, $|z| = D \sin|b|$. This choice of variables is convenient for identifying the galactic scale height of C_n^2. For example, a simple slab model with an exponential z dependence

$$C_n^2(z) = C_{n0}^2[1 - \exp(-|z|/H)], \qquad (5)$$

yields curves 1 through 4 in the figure that level off at $|z| \sim H$.

Open circles in Figure 3 correspond to $|b| < 10°$ and $DM > 50$ pc cm^{-3}, while filled circles correspond to $|b| \geq 10°$ or $DM \leq 50$ pc cm^{-3}. The notable features of the pulsar data in Figure 3 are:

- Most of the filled circles are segregated to small $SM_\perp$, as would be expected if higher latitude pulsars probe the A component of the medium.
- The filled circles show no sign of leveling off at high z and therefore provide no means of determining the scale height other than a lower limit, $H \lesssim 0.5$ kpc.

Also shown in Figure 3 are scattering diameters of extragalactic sources (after correcting for intrinsic sizes). Since lines of sight to these objects extend through

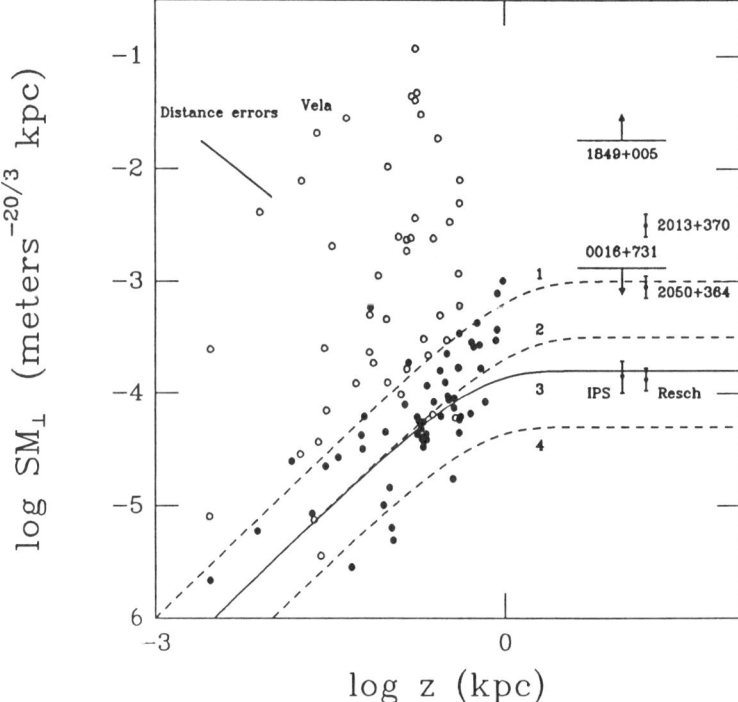

Figure 3 Perpendicular scattering measure $SM_\perp$ plotted against $|z|$ as derived from scintillation and VLBI measurements. For pulsars, filled circles are for $|b| \geq 10°$ or $DM \leq 50$ pc cm^{-3}; open circles are for $|b| < 10°$ and $DM > 50$ pc cm^{-3}. The slanted solid line shows the variation in $|z|$ and $SM_\perp$ due to a factor of two error in distance. The solid and dashed lines represent models (cf. eqn [5]) with parameters (1) $\log C_n^2 = -3$, $H = 1$ kpc; (2) $\log C_n^2 = -3.5$, $H = 1.0$ kpc; (3) $\log C_n^2 = -3.5$, $H = 0.5$ kpc; and (4) $\log C_n^2 = -4.0$, $H = 0.5$ kpc. VLBI measurements shown in the $z > 1$ kpc portion of the graph are scattering diameters or upper limits.

the entire scattering medium, they may be used to determine the scale height. The points marked 'IPS' and 'Resch' are combined results on large numbers of sources using interplanetary scintillations and two station VLBI at 70 MHz[7,8], respectively. The source 0016+731 is nearly unresolved[9] at a galactic latitude of 10.7°. The remaining sources are discussed in these proceedings (Mutel and Lestrade; Spangler and Cordes). A good representation of the filled circles is $\log C_{n0}^2 = -3.5$ and $H = 0.45 \pm 0.2$ kpc.

The distribution of C_n^2 in galactocentric radius R may also be constrained for the 'A' component. We assume an exponential distribution in z, as before, and a Gaussian in R with $1/e$ width A:

$$C_n^2(R, z) = C_n^2(GC) \exp(-|z|/H) \exp[-(R/A)^2]; \qquad (6)$$

$C_n^2(GC)$ is the value at the galactic center. VLBI on the extragalactic sources 0503+467 in the galactic anticenter direction[10] and 1909-161 in the first quadrant[11] respectively provide a lower limit $A \leq 7.5$ kpc and an upper limit $A \gtrsim 6$ kpc. We take the net $1/e$

radius to be $A = 7 \pm 2$ kpc. Details of this analysis may be found in references 2 and 10.

THE CLUMPED, INTENSE TURBULENCE COMPONENT

Sources at galactic latitudes $|b| \lesssim 1°$ show larger mean and larger scatter in $\overline{C_n^2}$ than high latitude sources. Ref 3 estimates the mean free path for intersecting a clump of intense turbulence to be $L_{mfp} \approx 5$ kpc. The associated volume filling factor for spherical clumps of size R is $f \approx R/L_{mfp} \approx 10^{-4} R_{pc}$. This estimate is obviously crude, since enhanced turbulence may occur in a heterogeneous mix of HII regions, supernova shells, and other regions. If clumps are of parsec size, the associated local values of $C_n^2 \approx \overline{C_n^2}(D/R)$ are 10^3 to 10^4 times larger than the maximum values shown in Figure 1.

For the assumed spectrum of eqn (1), the total rms electron density $\delta n_e(rms)$ for an 'outer scale' $\ell \equiv 2\pi/q_0 = \ell_{pc}$ pc is

$$\delta n_e(rms) \approx 1\, \ell_{pc}^{1/3} (C_n^2)^{1/2} \text{ cm}^{-3}$$

for C_n^2 in conventional units of $m^{-20/3}$. Clumps of parsec size may therefore require rms densities as large as 30 cm^{-3}.

Observational attempts[3,9,10,11,12,13,14] to link enhanced scattering to specific regions have not successfully associated it with a unique phase of the interstellar medium. It appears that supernova shocks and HII regions are both relevant.

This work was supported by the National Astronomy and Ionosphere Center at Cornell, which operates the Arecibo Observatory under contract to the National Science Foundation. This work was also supported by NSF grant 85-20530 to Cornell University and NSF grant 87-01979 to Carleton College.

REFERENCES

1. Rickett, B. J. 1977, *Ann. Rev. Ast. Ap.*, **15**, 479.
2. Cordes, J. M. and Spangler, S. R. 1988, submitted to *Ap. J.*
3. Cordes, J. M., Weisberg, J., and Boriakoff, V. 1985, *Ap. J.*, **288**, 221.
4. Lyne, A.G., Manchester, R. N., and Taylor, J. H. 1985, *M.N.R.A.S.*, **213**, 613.
5. Alurkar, S. K., Slee, O. B., and Bobra, A. D. 1986, *Aust J. Phys.* **39**, 433.
6. Clifton, T. R. and Lyne, A. G. 1986, *Nature*, **320**, 43.
7. Resch, G. 1974, Ph.D. Thesis, University of Florida.
8. Duffet-Smith, P. and Readhead, A. 1976, *M.N.R.A.S.*, **174**, 7.
9. Spangler, S. R., Mutel, R. L., Benson, J. M., and Cordes, J. M., 1986, *Ap. J.*, **301**, 312.
10. Spangler, S. R., Fey, A., and Cordes, J. M. 1987, *Ap. J.*, **322**, 909.
11. Dennison, B. *et al.* 1984, *Ast. Ap.*, **135**, 199.
12. Anantharamaiah, K. R. and Narayan, R., these proceedings.
13. Spangler, S. R. and Cordes, J. M., these proceedings.
14. Frail, D. A. and Clifton, T. R. 1988, *Ap. J.*, submitted.

THE ORIGIN OF SCATTERING IN THE INNER GALAXY

K. R. Anantharamaiah*, NRAO, Socorro NM 87801

Ramesh Narayan, Steward Observatory, Univ. of Arizona, Tucson AZ 85721

ABSTRACT

We argue that the enhanced scattering of radio waves observed at low galactic longitude and latitude is caused by the low-density outer envelopes of HII regions in the inner 8 kpc of the Galaxy. We develop a simple phenomenological theory to estimate the strength of density fluctuations (C_N^2) in a turbulent medium and apply the theory to the HII envelopes, using the densities and sizes determined from a recent 325 MHz recombination line survey. The predicted values of C_N^2 are in good agreement with the observations.

1. INTRODUCTION

Several observations of interstellar scintillation of pulsar signals and angular broadening of compact extragalactic radio sources suggest that electron density fluctuations in the ISM are significantly enhanced in the inner galaxy ($|l| \lesssim 30°, |b| \lesssim 2°$) compared to the solar neighborhood (Rao and Ananthakrishnan 1984, Dennison et al. 1984, Cordes, Weisberg and Boriakoff 1985). A measure of the magnitude of electron density fluctuations is the quantity, C_N^2; this is defined with respect to a power-law form of the three-dimensional power spectrum of spatial fluctuations (cf Cordes et al. 1985),

$$P_{3N}(\vec{q}) = C_N^2 \, |\vec{q}|^{-\beta}. \qquad (1)$$

Kolmogorov turbulence corresponds to $\beta = 11/3$; for this case, C_N^2 has units of $m^{-20/3}$.

It is known that the value of C_N^2 varies significantly over different regions of the ISM. The most detailed study has been that of Cordes et al. (1985), who measured the average C_N^2 along the lines-of-sight to a large number of pulsars. They find $C_N^2 \sim 3 \times 10^{-4}$ $m^{-20/3}$ in the solar vicinity, while for lines-of-sight longer than a few kpc towards the inner Galaxy they obtain $C_N^2 \sim 0.1 - 1$ $m^{-20/3}$. From a least squares fit to their data, Narayan (1987) estimates that $C_N^2 \sim 2$ $m^{-20/3}$ for lines-of-sight through the galactic plane with $|l| \lesssim 30°$, which is $\sim 10^4$ times the value for the local ISM. Even before these studies,

---

* On leave from Raman Research Institute, Bangalore 560080, India.

Dennison et al. (1984) had argued for enhanced scattering in the inner Galaxy from the absence of VLBI structure in compact extragalactic sources at low $|l|$, $|b|$. Similarly, Rao and Ananthakrishnan (1984) had noted an absence of IPS sources at 327 MHz at low $|l|$, $|b|$ and attributed this to enhanced scattering, with $C_N^2 \sim 1.5$ m$^{-20/3}$.

Cordes et al. (1985) suggested a two-component model for the electron density fluctuations in the Galaxy consisting of:

(a) A clumped, highly-scattering component with a scale height ≤ 100 pc and filling factor $\sim 10^{-4} - 10^{-2}$, indicative of association with extreme Population I material (HII regions, stellar wind bubbles, supernova remnants, etc.).

(b) A nearly uniform, moderately-scattering component with a scale height $\geq$ 500 pc and a large filling factor (> 0.5).

The enhanced scattering in the inner Galaxy is clearly associated with component (a). Rao and Ananthakrishnan (1984) and Dennison et al. (1984) suggested that this scattering may be caused by the ionized regions responsible for "diffuse" recombination line emission, also refered to as galactic ridge recombination lines (GRRLs) The GRRL emission was first observed by Gottesman and Gordon (1970) and has since been studied by several workers. Anantharamaiah (1985, 1986) deduced the parameters (densities, sizes) of the GRRL regions from a 325 MHz recombination line survey of the inner Galaxy.

In this paper, we develop a simple theory to calculate C_N^2 in a turbulent medium in terms of the mean density of the medium and the outer scale of the turbulence. We then use this theory to predict the value of C_N^2 in the GRRL regions, using Anantharamaiah's parameters for these regions. The predictions are in good agreement with the observed values of C_N^2 for the inner Galaxy, thus confirming the suggestion that the GRRL regions are responsible for the enhanced scattering.

2. RESULTS FROM THE 325 MHZ RECOMBINATION LINE SURVEY

Anantharamaiah (1985) combined his measured recombination line intensities at 325 MHz with those at higher frequencies (Hart and Pedlar 1976, Lockman 1976) to determine the electron densities of the regions producing the GRRLs. The estimated densities are $\sim 1 - 10$ cm^{-3}, and the sizes of the regions are $\sim 20 - 200$ pc, based on the geometry of the line-emitting regions and their continuum emission. Furthermore, the close correlation between the velocities of the 325 MHz recombination lines and those of HII regions in the same direction strongly indicates that the GRRL regions are low-density outer envelopes of standard HII regions (Anantharamaiah 1986).

HII regions are known to be numerous in the inner Galaxy, peaking over the galactocentric radius range 4-8 kpc; they have typical sizes in the range 1-10 pc and electron densities of 10^2-10^4 cm^{-3}. If each HII region had an envelope of ~ 100 pc, then almost every line of sight with $|l| \lesssim 40°$ and $|b| \lesssim 1°$ would intersect at least one such envelope. These envelopes thus constitute a population with the right galactic distribution and covering factor to explain the enhanced C_N^2 observed towards the inner Galaxy. It remains to show that they have the necessary density fluctuations and we turn to this next.

3. ESTIMATING C_N^2

The structure function of density fluctuations, $D_N(\vec{r})$, is defined to be

$$D_N(\vec{r}) \equiv \langle [N_e(\vec{r}' + \vec{r}) - N_e(\vec{r}')]^2 \rangle, \tag{2}$$

where $N_e(\vec{r})$ is the electron density at position $\vec{r}$. The density spectrum (1) with $\beta = 11/3$ (for Kolmogorov turbulence) gives

$$D_N(\vec{r}) = 30.30\ C_N^2 |\vec{r}|^{2/3}. \tag{3}$$

We now make the simplifying assumption that the electron density fluctuations are strongly non-linear on the outer scale of the turbulence, r_{out}, and describe the degree of non-linearity by a parameter, α, such that the rms density fluctuations on the scale r_{out} is $\alpha \bar{N}_e$, where $\bar{N}_e$ is the mean electron density of the medium. This is a purely phenomenological approach to estimating C_N^2 and, at this time, we do not claim that the parameter α has any particular theoretical significance. (It could be a measure of true non-linearity or may represent merely some sort of a clumping factor.) However, on simple considerations we do expect α to be greater than or of the order of unity. Below, we estimate the value of α "experimentally" by considering regions with known scattering properties. With our assumptions, eq. (3) can be rewrittten as

$$C_N^2 \sim 0.066\ \alpha^2 \bar{N}_e^2 r_{\text{out}}^{-2/3}. \tag{4}$$

If $\bar{N}_e$ is in m^{-3} and r_{out} in m, then C_N^2 has the standard units of m$^{-20/3}$.

The estimate of C_N^2 given in eq. (4) can be compared directly with observations whenever the scattering is uniform along the line-of-sight. However, the scattering is often localized to a small volume and the obervations give only a mean C_N^2. If the scattering is restricted to a path-length L, then eq. (4) needs to be replaced by

$$C_N^2 = 0.066\ \alpha^2 \bar{N}_e^2 r_{\text{out}}^{-2/3} \left[3 \frac{L}{D_{\text{eff}}} \left(\frac{D_1}{D} \right)^2 \right]. \tag{5}$$

Here D is the total distance from the observer to the source (infinity for extragalactic sources), D_1 is the distance from the scattering screen to the source, D_{eff} is the effective distance through the galactic disk that was assumed in calculating the "measured" C_N^2, and the factor of 3 is for a particular assumption regarding the geometry. We now apply eqs. (4) and (5) to various regions of scattering in the ISM.

Local galactic disk: Using pulsar dispersion measures it has been estimated that the mean electron density of the ISM in the solar vicinity is $\bar{N}_e \sim 0.025$ cm^{-3} (e.g. Lyne, Manchester and Taylor 1985). This medium appears to be relatively smooth and is most probably the site of component (b) of Cordes *et al.* (1985). If we set r_{out} equal to the scale height of this component, which is ~ 500 pc, eq. (4) gives $C_N^2 \sim 6.6 \times 10^{-6} \alpha^2$ m$^{-20/3}$. Comparing with the measured value of $\sim 3 \times 10^{-4}$ m$^{-20/3}$, we find that we need $\alpha \sim 7$ in our phenomenological theory. It is reassuring that α is greater than unity, but not grossly so.

Vela pulsar: Even though the Vela pulsar is nearby ($D = D_{\text{eff}} = 500$ pc), the line-of-sight to this pulsar gives a measured C_N^2 of 1.4 m$^{-20/3}$, which is much in excess of that for other local lines-of-sight. The reason for the enhanced scattering seems to be the Gum nebula, a large shell of low-density ionized gas. Combining Reynolds' (1976) emission measure of the Gum nebula in the direction of the Vela pulsar with the dispersion measure of the pulsar (most of which is due to the nebula), we estimate $\bar{N}_e = 4.5$ cm^{-3} and $L = 12$ pc for the layer in the nebula that does most of the scattering. If we assume $D_1 \sim 125$ pc (radius of the Gum nebula shell), and $r_{\text{out}} = L$, eq. (5) gives $C_N^2 = 1.2 \times 10^{-2} \alpha^2$ m$^{-20/3}$. Comparing with the measured value of C_N^2 given above, we find that we need $\alpha \sim 11$, which is similar to the value of α we had obtained from the local ISM. Considering the very large difference in the level of scattering in the two cases (a factor $\sim 10^4$ in C_N^2), this agreement is quite encouraging and suggests that our simple-minded scheme does have some merit. We use an average α of 10 in the calculation described below.

Low-density Envelopes of HII regions: As discussed earlier, the mean electron density of these regions is ~ 5 cm^{-3} and their mean size is $L \sim r_{\text{out}} \sim 100$ pc. We take D_{eff} to be the total pathlength through the galactic disk, ~ 20 kpc, and $D = D_1 = \infty$ (for an extragalactic radio source). Using eq. (5) we then get $C_N^2 = 1.2$ m$^{-20/3}$. This is in very good agreement with the measured values quoted in sec. 1, leading us to believe that the GRRL-producing envelopes of HII regions are indeed the source of the very strong scattering found towards the inner Galaxy. Note that the parameters we used are averages and any particular line-of-sight could fluctuate around these values.

4. ANGULAR BROADENING AT 325 MHz

Blandford and Narayan (1985) and Romani, Narayan and Blandford (1986) show that the average scattering-radius of an extragalactic radio source due to interstellar scattering is

$$\theta_{scatt} \sim 1 \ (C_N^2)^{3/5} \lambda_m^{11/5} D^{3/5} \text{ arcsec,} \qquad (6)$$

where λ_m is the wavelength in meters and D is the pathlength through the ISM in kpc. As discussed in sec. 1, the mean C_N^2 for lines of sight at low galactic latitudes with $|l| \lesssim 30°$ is ~ 1.5 m$^{-20/3}$. Substituting this value into eq. (6), the mean scattering-diameter of compact extragalactic radio sources at low latitudes and longitudes is expected to be $2\theta_{scatt} \sim 15 \lambda_m^{11/5}$ arcsec.

At 325 MHz ($\lambda_m = 91$ cm) the expected angular diameter is ~ 12 arcsec, which is resolvable with the Very Large Array of the National Radio Astronomy Observatory. We are currently making a systematic search to verify this large predicted angular broadening. It is our hope that this will ultimately develop into a useful probe of low-density ionized regions in the inner Galaxy.

Acknowledgements: This work was supported in part by NASA Astrophysical Theory grant NAGW-763. The National Radio Astronomy Observatory is operated by the Associated Universities Inc., under contract with the National Science Foundation.

REFERENCES

Anantharamaiah, K.R., 1985. *J. Astrophys. Astr.*, **6**, 177, 203.
Anantharamaiah, K.R., 1986. *J. Astrophys. Astr.*, **7**, 131.
Blandford, R. and Narayan, R., 1985. *Mon. Not. R. astr. Soc.*, **213**, 591.
Cordes, J.M., Weisberg, J.M. and Boriakoff, V., 1985. *Astrophys. J.*, **288**, 221.
Dennison, B., Thomas, M., Booth, R.S., Brown, R.L., Broderick, J.J. and Condon, J.J., 1984. *Astr. Astrophys.*, **135**, 199.
Gottesman, S.T. and Gordon, M.A., 1970. *Astrophys. J.*, **162**, L93.
Hart, L. and Pedlar, A., 1976. *Mon. Not. R. astr. Soc.*, **176**, 135, 547.
Lockman, F.J., 1976. *Astrophys. J.*, **209**, 429.
Lyne, A.G., Manchester, R.N. and Taylor, J.H., 1985. *Mon. Not. R. astr. Soc.*, **213**, 613.
Narayan, R., 1987. *Astrophys. J.*, **319**, 162.
Rao, A.P. and Ananthakrishnan, S., 1984. *Nature*, **312**, 707.
Reynolds, R.J., 1976. *Astrophys. J.*, **203**, 151.
Romani, R.W., Narayan, R. and Blandford, R., 1986. *Mon. Not. R. astr. Soc.*, **220**, 19.

VLBI ANGULAR BROADENING MEASUREMENTS IN THE CYGNUS REGION

A. L. Fey, S. R. Spangler, and R. L. Mutel
The University of Iowa, Iowa City, Iowa 52242

ABSTRACT

We report multi-frequency VLBI observations of angular broadening of eight compact radio sources whose lines of sight pass through the galactic plane in Cygnus. We use the characteristic λ^2 dependence of angular size to distinguish between intrinsic and scattered structure. Four of the eight sources have measured angular sizes larger than would be expected for sources with such spectra and exhibit a λ^2 dependence of angular size. These results lead us to the conclusion that we have observed angular broadening for these four sources. The structures of the other four sources do not appear to be dominated by interstellar scattering. Our results indicate that scattering in the Cygnus region changes by a factor of 2 − 5 for sources separated by only a couple of degrees. With the exception of one of the sources, the observed angular broadening is confined within a few degrees of the galactic plane.

INTRODUCTION

We have observed eight compact extragalactic radio sources whose lines of sight pass through an extended X-ray source commonly referred to as the Cygnus superbubble[1]. This region, in the constellation Cygnus ($\ell = 65° - 90°, |b| \leq 10°$), contains numerous supernova remnants and OB-associations and therefore many likely candidates for the generation of electron density turbulence. The objective of these observations is to determine the distribution of scattering material in the Cygnus region.

Since the point source visibility function is directly related to the turbulent electron density power spectrum[2], angular broadening measurements provide a powerful way to determine properties of interstellar turbulence. In addition, the magnitude of angular broadening can provide a measurement of the path integral of the electron density variance.

Previous estimates of scattering in the Cygnus region have been made mostly by pulsar pulse broadening measurements. Due to the uncertainty in pulsar distance estimates, it is uncertain how reliable it is to compare scattering measurements along two different lines of sight when one is determined by pulsar pulse broadening and the other by angular broadening. The primary motivation for the present observations was, therefore, to obtain a consistent set of scattering estimates so that we could reliably compare scattering strength along different lines of sight in the Cygnus region.

OBSERVATIONS

We have made observations of eight sources at frequencies of 0.61, 1.66, and 4.99 GHz using Mark II VLBI. The source list is indicated in Table 1. The 4.99 GHz observations were carried out in October 1985 using antennas at Bonn, Haystack, Fort Davis, Owens Valley, and Hat Creek. The 18 cm VLB interferometer consisted of Owens Valley, NRAO-Green Bank, Haystack, Hat Creek, Onsala, and the phased VLA. The

18 cm observations were done in March 1986. Finally, 50 cm observations were carried out in October 1986 using antennas at Jodrell Bank, Westerbork, Iowa, NRAO-Green Bank, Bonn, and Owens Valley. The observations at each frequency were made using short scans (7.5 minutes for 6 and 18 cm; 15 minutes for 50 cm) over a number of different hour angles to maximize the (u,v) coverage.

ANALYSIS

A source whose structure is dominated by angular broadening should show a measured angular size that scales with frequency as[2]

$$\theta_S \propto \nu^\gamma, \quad \gamma \approx -2, \tag{1}$$

where the exact value of γ depends on the spatial spectrum of the turbulent electron density fluctuations. To get an estimate of θ_{1GHz}, the scattered angular size at a frequency of 1 GHz, we do a least squares fit of equation (1) to the measured angular sizes. The values for the four sources whose structures are dominated by scattering, 2005+403, 2021+317, 2023+336, and 2048+313, are listed in Table 1.

Table 1
Scattering Strength Measurements

Source	ℓ (Deg)	b (Deg)	Angular Size $\theta_{1GHz}(mas)$	Index γ	Scattering Measure $SM\ (m^{-20/3}kpc)$
(1)	(2)	(3)	(4)	(5)	(6)
1923+210	55.6	2.3	≤ 16.3		$\leq 3.2 \times 10^{-2}$
1954+513	85.3	11.8	≤ 4.3		$\leq 3.5 \times 10^{-3}$
2005+403	76.8	4.3	79.5 ± 10.4	-1.94 ± 0.15	$(4.5 \pm 1.0) \times 10^{-1}$
2013+370[3]	74.9	1.2	46.4 ± 9.6	-2.0 ± 0.2	$(1.8 \pm 0.6) \times 10^{-1}$
2021+317	71.4	-3.1	25.7 ± 2.6	-2.08 ± 0.13	$(6.8 \pm 1.2) \times 10^{-2}$
2022+542	90.1	9.7	≤ 1.4		$\leq 5.3 \times 10^{-4}$
2023+336	73.1	-2.4	67.6 ± 14.3	-1.97 ± 0.29	$(3.4 \pm 1.2) \times 10^{-1}$
2048+313	74.6	-8.0	64.7 ± 11.0	-2.19 ± 0.18	$(3.2 \pm 0.9) \times 10^{-1}$
2050+364[4]	78.9	-5.1	9.3 ± 2.5	-1.8 ± 0.2	$(1.3 \pm 0.6) \times 10^{-2}$
2113+293	76.6	-13.3	≤ 1.2		$\leq 4.1 \times 10^{-4}$
3C418[3]	88.8	6.0	≤ 8.0		$\leq 9.8 \times 10^{-3}$

Column (1) gives the source name. Columns (2) and (3) list the galactic longitude and latitude of each source, respectively. The values of θ_{1GHz} are listed in column (4). Column (5) lists the values of the index γ and finally column (6) gives an estimate of the scattering strength which will be discussed below. We also include in Table 1 measurements from the literature for three other scattered sources: 2013+370, 3C418, and 2050+364.

If scattering does not dominate the observed source structure, we would like to obtain a reasonable estimate of the scattering contribution to the observed structure.

We note that the observed structure is a convolution of the intrinsic structure with a scattering function. We follow an analysis formulated by Spangler et al.[5] and assume that both the intrinsic and scattered structure are of Gaussian form.

At this point we note that the measured flux density spectral indices for the four sources 1923+210, 1954+513, 2022+542, and 2113+293 are 0.14, −0.07, −0.01, and −0.10, respectively. The intrinsic angular size of a flat spectrum, inhomogeneous synchrotron source should scale roughly as the inverse of the observing frequency[6]

$$\theta_I \propto \nu^\gamma, \ \gamma \approx -1. \tag{2}$$

We assume equation (2) describes the intrinsic structure of these four sources and use equation (1) above for the scattered structure. We then have for the observed angular size

$$\theta^2_{observed} = \frac{\theta^2_{I_o}}{\nu^2} + \frac{\theta^2_{S_o}}{\nu^4}, \tag{3}$$

where ν is the observing frequency in GHz and θ_{I_o} and θ_{S_o} are the intrinsic and scattered angular sizes scaled to a frequency of 1 GHz. Multi-frequency observations allow us to fit for the two unknowns in this equation. The estimates of θ_{S_o} for the four sources 1923+210, 1954+513, 2022+542, and 2113+293 are listed in column (4) of Table 1 as θ_{1GHz}.

We are interested primarily with the variation in the magnitude of scattering along different lines of sight. We define a scattering measure as the line of sight integral of the normalization constant of an assumed power law power spectrum, and adopt a value for the power law index of 11/3 (the Kolmogorov value). From Cordes and Spangler[7] we have for the scattering measure

$$SM = \int_0^L C_N^2(s)ds = 3.05 \times 10^{-4} \theta_{1GHz}^{5/3} \ m^{-20/3}kpc, \tag{4}$$

where θ_{1GHz} is in milliarcsec. The path length through the scattering medium is designated by L and is in kiloparsecs. The values of SM for all eight sources are listed in Table 1.

RESULTS

Figure 1 shows the main results of this paper in the form of a plot of scattering measure (SM) versus galactic longitude and latitude.

Galactic Distribution

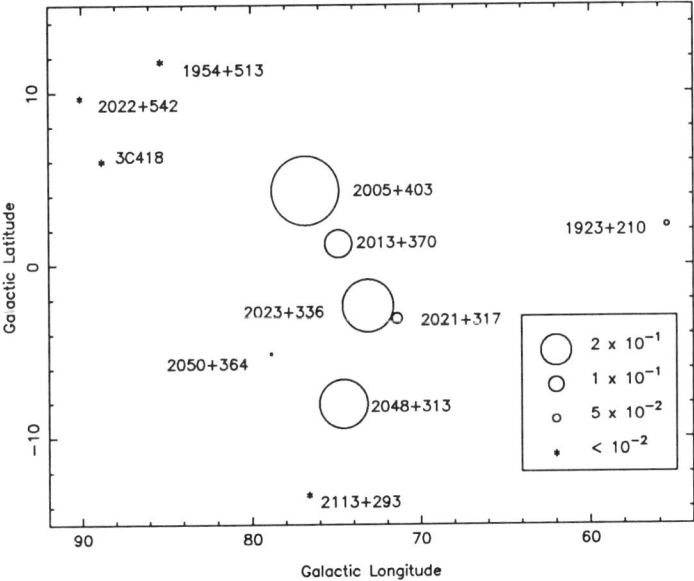

Figure 1: Scattering measure (SM) versus galactic longitude and latitude.

The sizes of the open circles are directly proportional to the scattering measure as indicated by the key. We have also included in this figure values for the sources 2013+370[3], 3C418[3] and 2050+364[4]. The important thing to note is the factor of five change in SM over $\approx 2°$ between the sources 2021+317 and 2023+336. With an assumed distance of two kiloparsecs to the Cygnus "Superbubble", this represents a spatial scale of $\approx 70\,pc$. Also note the factor of two and one half difference in SM over $\approx 4°$ ($\approx 140\,pc$) between the sources 2005+403 and 2013+370. The scattered sources are confined to within a couple of degrees of the plane, with the exception of 2048+313, whose line of sight passes through the Cygnus Loop supernova remnant.

CONCLUSIONS

1) We have observed angular broadening for the following sources: 2005+403, 2021+317, 2023+336, and 2048+313.

2) The structures of the remaining sources, 1923+210, 1954+513, 2022+542, and 2113+293 do not appear to be dominated by interstellar scattering. It is possible, however, that scattering becomes important for 1923+210 at wavelengths larger than 18 cm.

3) The scattering measure (defined as the line of sight integral of the normalization constant of the turbulent electron density power spectrum) in the Cygnus region changes by a factor of 2 − 5 for sources separated by only a couple of degrees.

4) With the exception of 2048+313, the observed angular broadening is confined to within several degrees of the galactic plane.

ACKNOWLEDGEMENTS

This research was supported at the University of Iowa by grants NAGW-806 and 831 from the National Aeronautics and Space Administration.

REFERENCES

(1) Bochkarev, N. G., Sitnik, T. G., *Ap. Space Sci.*, **108**, 237. (1985)

(2) Rickett, B. J., *Ann. Rev. Astr. Ap.*, **15**, 479. (1977)

(3) Spangler, S. R., Mutel, R. L., Benson, J. M., Cordes, J. M., *Ap. J.*, **301**, 312. (1986)

(4) Mutel, R. L., and Hodges, M. W., *Ap. J.*, **307**, 472. (1986)

(5) Spangler, S. R., Fey A. L., and Cordes, J. M., *Ap. J.*, **322**, 909. (1987)

(6) Kellermann, K. I., and Pauliny-Toth, I. I. K., *Ann. Rev. Astr. Ap.*, **19**, 373. (1981)

(7) Cordes, J. M., and Spangler, S. R., submitted to *Ap. J.* (1987)

Preliminary Results from a 7 Station VLBI Survey of OH Masers in the Galactic Plane

P. J. Diamond (NRAO), A. Martinson (Onsala),
B. Dennison (VPI), R. S. Booth (Onsala), A. Winnberg (Onsala)

Abstract

An initial analysis of a large VLBI survey of OH masers in the Galactic Plane reveals a correlation of maser component angular size with source distance. This correlation is presumably due to the effect of an intervening scattering medium.

Introduction

Observations of compact extragalactic radio sources[1] and pulsars[2] have demonstrated that the ionized gas responsible for interstellar scattering is distributed in a complex manner with at least two components, one of high scale height ($\sim$ 1kpc) and appreciable filling factor, and another having low scale height ($\sim$ 100pc), low filling factor ($\sim 10^{-4}$) and probably a complex Galactic distribution. The phenomenology with which this latter component is associated is unknown. With the aim of attacking this particular problem we decided to perform a high spatial and frequency resolution survey of $\sim$ 60 Galactic Plane OH masers. Such observations will provide scattering measurements over a broad range of Galactic longitudes and distances.

Observations and Results

In April 1985 we spent 48 hours observing our maser sample using a 7-station VLBI array (Effelsberg 100m; Westerbork 25m; Onsala 26m; Jodrell Bank 76m; Haystack 37m; Green Bank 40m; Owens Valley 37m). About 30 sources (mainly OH/IR stars) were observed in the OH satellite line frequency of 1612.231 MHz, the remainder (mainly OH sources associated with regions of star formation) were observed at the OH main-line frequencies of 1665.4018 MHz and 1667.359 MHz. Here we report our initial analysis of 12 sources observed at 1665 MHz. Each source was observed for at least two 15 minute periods during its time above the horizon. The observation times were chosen in order to maximise the spatial frequency sampling for each source. Depending on the velocity width of the individual sources an observing bandwidth of either 250 kHz of 62.5 Khz was employed.

The data were correlated on the MkII correlator of the Max Planck Institut für Radioastronomie in Bonn, West Germany. The correlator was run in its normal spectral line mode which generates 128 channel cross-power spectra, the frequency resolution was therefore 2.3 kHz for the broader band observations and 0.6 kHz for the narrower band. The antenna gains were calibrated using the autocorrelation spectra generated in the correlation procedure. The delays were calibrated by frequent observations of compact continuum sources interspersed in our schedule; the fringe rates were calibrated seperately for each source and the data then averaged in time and edited.

We then attempted to fit single Gaussian models to the visibilities of the stongest channel (or averaged groups of channels) for each source. In most cases examination of the visibilities demonstrated this process to be a valid one since the structure in the individual channels was, on the whole, simple. However in one or two cases the visibilities were obviously complex and the results may therefore not be reliable. We intend to resolve this problem in the future by more intelligent analysis of the data. An additional uncertainty in our results is that we estimated the angular size for the strongest maser feature only, it is probable that there is some intrinsic variation of maser size within the source[3], however we do not believe this effect will change our principal results.

In Table 1 we have listed the results for the sources we detected at 1665 MHz. One problem evident in the interpretation of the results is that in a few sources the correct distances are unknown. When this is the case we have listed the near and far kinematic distances with the least favoured value in parentheses. The results are presented graphically in Fig. 1 where we have plotted the OH maser angular size against distance; it is obvious that a correlation exists in the sense that the further away a maser is the larger its angular size. This has always been suggested by previous VLBI observations but Fig. 1 is the first real confirmation that this is true.

TABLE 1
SOURCE COMPONENT SIZES

Source	Gal. Long.	Gal. Lat.	Distance kpc	Θ mas
W33A	12.91	−0.26	4.0	20.0 ± 3.0
OH20.1	20.08	−0.13	7.7(4.0)	35.0 ± 2.0
OH35.2	35.20	−1.73	2.3	3.0 ± 0.5
OH40.6	40.62	−0.14	2.1(13.1)	14.0 ± 2.0
OH43.8	43.79	−0.13	2.7(11.8)	7.6 ± 0.5
OH45.5	45.47	0.05	9.7	3.0 ± 1.0
SCrB	49.47	57.17	0.375	13.0 ± 10.0
W75S	81.72	0.57	2.5	4.0 ± 0.3
Cep A	109.87	2.11	0.73	3.0 ± 0.7
NGC7538	111.537	0.77	3.5	13.0 ± 1.0
W3(OH)	133.95	1.06	2.2	6.0 ± 3.0
Mon R2	213.72	−12.59	1.0	2.0 ± 0.5

The correlation does not hold for all sources. In Fig. 1 the source with an angular size of 100 mas is the OH maser in Sgr B2[4], this particular source is presumably highly scattered since it lies so close to the Galactic Centre. Two other sources have very small sizes for their distance, these are both at Galactic Longitude > 45° and may well be seen through a hole in the scattering screen.

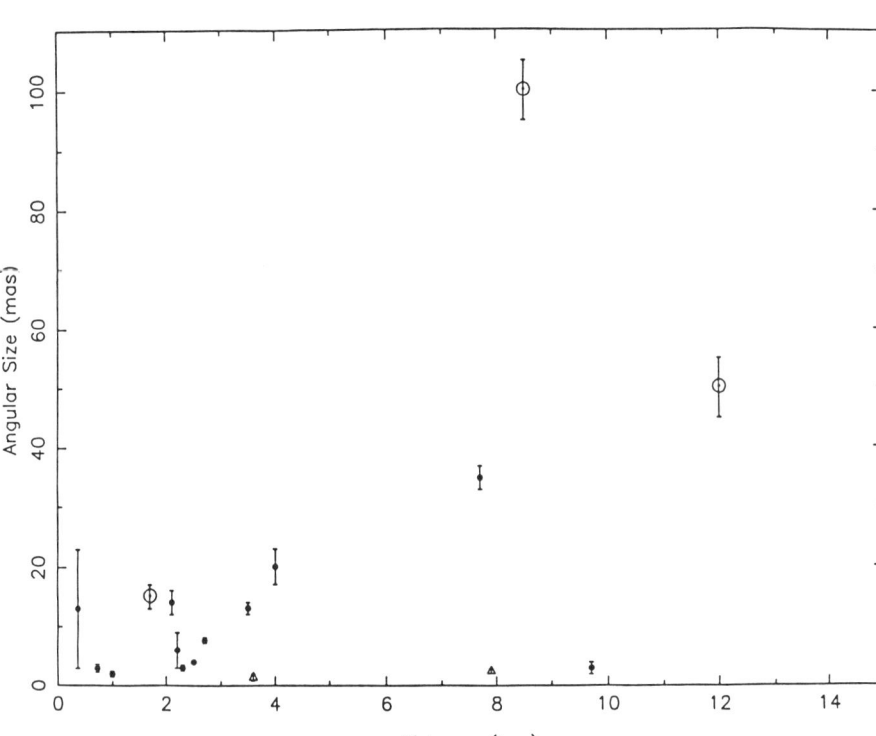

Fig. 1. A plot of the angular sizes of the OH masers against distance from the Sun. The solid circles are from this paper; the open circles are from Kent and Mutel (1982) and Burke et al. (1968); the open triangles are from Kemball et al. (1988).

The average scattering radius of an extragalactic source due to interstellar scattering is[5]

$$\Theta_{scatt} \sim 1(C_N^2)^{3/5} \lambda_m^{11/5} D^{3/5} arcsec$$

where λ_m is the observing wavelength (metres) and D is the pathlength through the ISM (kpc), in our case we assume D is the distance to the source. A least squares fit to our data demonstrate that $\Theta_{scatt} \propto D^{0.4\pm0.3}$, this is in general agreement with the theoretical result. In addition we estimate C_N^2, the strength of the electron density fluctuations in the ISM, to be $\sim 0.25 \pm 0.10 m^{-20/3}$, in agreement with values determined by others[2].

Conclusions

We have performed a VLBI survey of OH masers in the Galactic Plane and initial analysis of the data reveals a correlation of component angular size with distance for most sources. Some sources do not fit the simple picture and may reflect the inhomogeneity of the scattering medium.

References

1. Dennison, B. *et al.*, 1984, *Astr. Ap.*, **135**, 199
2. Cordes, J. M., Weisberg, J. M. and Boriakoff, V., 1985, *Ap. J.*, **288**, 221.
3. Gwinn, C. R., Moran, J. M., Reid, M. J. and Schneps, M. H., 1988, *Ap. J.*, in press.
4. Moran, J. M., 1968, Ph. D Thesis, Massachusetts Institute of Technology.
5. Blandford, R. and Naryan, R., 1985, *M.N.R.A.S.*, **213**, 591.
6. Kemball, A., Diamond, P. J. and Mantovani, F., 1988, *M.N.R.A.S.*, in press.
7. Kent, S. R. and Mutel, R. L., 1982, *Ap. J.*, **263**, 145.
8. Burke, B. F., *et al.*, 1968, *A.J.*, , **73**, 168.

VII. IMAGING TECHNIQUES AND PULSAR STUDIES

THE SHAPE OF A SCATTERED IMAGE—THEORY AND NUMERICAL SIMULATIONS

Jeremy Goodman,*
Institute for Advanced Study, Princeton NJ 08540

Ramesh Narayan
Steward Observatory, Univ. of Arizona, Tucson, AZ 85721

ABSTRACT

We have studied theoretically and numerically the scatter-broadened image of a radio point source in the presence of strong interstellar scintillation. The characteristics of the image depend on the integration time, t_{int}, of the observations. Normal VLBI corresponds to t_{int} lying between the diffractive timescale, t_{diff}, and the refractive timescale, t_{ref}. The shape of the image in this case is insensitive to the bandwidth employed but does depend on whether the power spectrum of density fluctuations in the ISM is shallow or steep (power law index < 4 or > 4). There is also some dependence on the inner scale of the spectrum. This suggests that VLBI imaging could be a useful diagnostic of the density fluctuation spectrum. If $t_{int} < t_{diff}$, and if the bandwidth of the observations is less than the decorrelation bandwidth of the intensity scintillations, the image has deeply modulated structure down to the resolution limit of the telescope.

1. INTRODUCTION

As these proceedings attest, the direct observational study of scatter-broadened radio images has begun[1-7]. The work summarized here was an attempt to understand how scatter-broadened images are formed and how their characteristics depend on the integration time t_{int} over which visibilities are accumulated in a VLBI experiment. A fuller account of this work will appear elsewhere[8].

Cornwell and collaborators[9,2] have developed similar ideas and have discussed the important differences between filled and unfilled apertures, a topic we have largely neglected.

The theory of optical image formation through a turbulent atmosphere is well developed[10], but much of that work cannot be directly applied to the radio case. In particular, the average image (see below) has no analog in optical astronomy. These and other differences stem from the fact that optical astronomy is done in the weak scintillation regime, whereas radio scintillation is very strong.

Our emphasis is conceptual and qualitative, but this meeting has demonstrated the need for more quantitative theoretical predictions, including the effects of finite observational bandwidth, if these new data are to be used to infer statistical properties of the scattering medium. Such work is now in progress.

2. SNAPSHOT, AVERAGE, AND ENSEMBLE-AVERAGE IMAGES

In this work, we assume a thin-screen model in which the scintillation pattern is fixed while the observer moves through the pattern at velocity v. Using v, we can relate

---
* Current address: Princeton Univ. Observatory, Princeton NJ 08544

Imaging Techniques

t_{int} to a lengthscale $r_{\text{int}} \equiv v t_{\text{int}}$ transverse to the Earth-pulsar line of sight, which is to be compared to the following scales:

$$\begin{aligned}
r_{\text{ref}} &\equiv D\theta_* = \quad \text{projected image size} \equiv \quad vt_{\text{ref}}, \\
r_F &\equiv \sqrt{\lambda D/2\pi} = \quad \text{Fresnel length}, \\
r_{\text{diff}} &\equiv r_F^2/r_{\text{ref}} \equiv \quad vt_{\text{diff}}, \\
r_b &\equiv r_F^2/b_{\text{max}} = \quad \text{projected beam size.}
\end{aligned} \qquad (1)$$

Here θ_* is the angular half-width of the image, λ is the radio wavelength, b_{max} is the longest baseline in the interferometer, t_{diff} is the diffractive scintillation timescale, and D is an effective distance to the scattering screen. If the wavefronts impinging on the scattering screen were planar, D would be the distance D_{OS} between the observer and the screen; for a pulsar at a finite distance D_{SP} behind the screen, $D^{-1} \equiv D_{\text{OS}}^{-1} + D_{\text{SP}}^{-1}$. We assume that $r_b < r_{\text{ref}}$, so that the image is resolved; this requires $b_{\text{max}} > r_{\text{diff}}$.

The scatter-broadened image is actually composed of a large number of *subimages*, whose projected locations $\mathbf{r}_i$ on the scattering screen are determined by Fermat's principle[11]:

$$\vec{\nabla}_r \left[\phi(\mathbf{r}) + \frac{(\mathbf{r} - \mathbf{R})^2}{2r_F^2}\right] = 0, \qquad (2)$$

in which the first term is the phase shift produced at $\mathbf{r}$ by the scattering layer, and $\mathbf{R}$ is the transverse position of the observer ($\mathbf{r}$ and $\mathbf{R}$ are two-vectors lying respectively in the scattering screen and in the parallel "observer's" plane). The subimages are concentrated in a region of projected width $\sim r_{\text{ref}}$. In strong scintillation, where $r_{\text{ref}} \gg r_F$, the total number of subimages, N, is very large. For a "Kolmogorov" or similar spectrum with ignorable inner and outer scales ("Type A" in Narayan's[12] taxonomy), $N \sim (r_{\text{ref}}/r_{\text{diff}})^2 = (D\theta_*/r_F)^4$. For good optical seeing, in contrast, the variation of ϕ across the Fresnel length is less than unity, so that geometric-optics subimages do not really exist.

Each of these subimages has a complex amplitude $A_i = \sqrt{F_i}\exp(i\phi_i)$, where $\phi_i \equiv \phi(\mathbf{r}_i)$ and F_i is the total flux carried by the ith subimage. If $t_{\text{int}} \ll t_{\text{diff}}$ and the observational bandwidth is smaller that the decorrelation bandwidth $\Delta\nu_{\text{dc}} \equiv 2c/D\theta_*^2$, then the subimages within a resolution element of projected width r_b add *coherently*; that is, the total flux of that element is the absolute square of the sum of the complex amplitudes. This is what we call the "snapshot" image.

The snapshot image is deeply modulated on all scales larger than r_b. For a Type A spectrum, the A_i are approximately independent, so that the flux F_b within a resolution element is the sum of a large number of independent, identically-distributed complex random variables and therefore follows a Rayleigh distribution, $p(F_b)dF_b = \exp(-F_b/\bar{F}_b)dF_b$, which has a variance equal to its mean. The flux in one resolution element is independent of the flux in a neighboring element, apart from a decrease in the beam flux towards the edge of the image. Hence, the snapshot image looks like a two-dimensional white-noise process that has been smoothed over a length $\approx r_b$ and multiplied by an envelope or window function of width $\approx r_{\text{ref}}$. Versions of the snapshot image made with differing interferometric resolutions r_b bear no simple relation to one another, even if the observations are made simultaneously and over the same bandwidth. In particular, the lower-resolution image does not look like a smoothed version of the higher-resolution one, because the subimage amplitudes are combined differently in the two cases and can sum to unrelated beam fluxes.

The coherence of the subimages depends on the properties of the source as well as the observing parameters. If the intrinsic angular size of the source, θ_0, is larger than λ/r_b, then the coherence is lost for the same reason that sources with $\theta_0 > \lambda/r_{\text{ref}}$ do not scintillate. The only difference is that the snapshot image depends on phase relations among subimages falling within the same resolution element, whereas scintillation involves the phase relations among *all* of the subimages.

If two subimages are separated by an angle $\theta_b \equiv r_b/D$, then their relative phase measured on the observer's plane varies cyclically with a repeat wavelength of $\lambda/\theta_b = b_{\max}$. Thus, as the interferometer traverses its own width through the scintillation pattern, the phase relations among most of the subimages within a given resolution element change by a few radians. Consequently, the details of the snapshot image decorrelate over a time $\sim b_{\max}/v$. When visibilities are accumulated over an integration time t_{int} long compared to $b_{\max}/v$ but short compared to r_b/v, the individual subimages persist and maintain roughly constant fluxes, but their amplitudes add *incoherently*; that is, the flux within the beam is the sum of the fluxes of the subimages within the beam. The result of such an experiment is what we call the "average" image.

Another way to get the average image is to use a bandwidth larger than $\Delta\nu_b$, defined as follows. Consider two subimages at positions $\mathbf{r}_1$ and $\mathbf{r}_2$. The difference in the path lengths between the observer and the two of them is $\Delta D = (\mathbf{r}_1 - \mathbf{r}_2) \cdot (\mathbf{r}_1 + \mathbf{r}_2)/2D$. If the two subimages fall within the same resolution element, and that element is at a typical location within the image, then $|\mathbf{r}_1 - \mathbf{r}_2| \sim r_b$ and $|\mathbf{r}_1 + \mathbf{r}_2|/2 \sim r_{\text{ref}}$, so $\Delta D \sim r_b r_{\text{ref}}/D$. The relative phase between them changes by 2π radians when the frequency changes by $\Delta\nu_b = c/\Delta D \sim cD/r_{\text{ref}} r_b = (r_{\text{ref}}/r_b)\Delta\nu_{\text{dc}}$.

Finally, if t_{int} is very long, then individual subimages will have been created and destroyed (in pairs) many times with varying fluxes and varying positions within the image. The result of such a long-term average is the "ensemble-average" image, so-called because it should be equivalent to the average over an ensemble of scattering screens with different realizations of the random phase function $\phi(\mathbf{r})$. The ensemble-average image depends only on the statistics of ϕ. In fact, if $D_\phi(r) \equiv \langle [\phi(\mathbf{r}') - \phi(\mathbf{r}' + \mathbf{r})]^2 \rangle$ is the structure function of the phase, then the ensemble-averaged visibility on a baseline $\mathbf{b}$, $\langle V(\mathbf{b}) \rangle$, is[13]

$$\langle V(\mathbf{b}) \rangle = \exp\left[-\frac{1}{2}D_\phi(b)\right]. \quad (3)$$

The ensemble-average image is the Fourier transform of (3). Since $\langle V(\mathbf{b}) \rangle$ falls exponentially at large baselines, the ensemble-average image is smooth and structureless on scales $\ll r_{\text{ref}}$.

The following table summarizes the three kinds of image and the ranges of t_{int} for which they apply, assuming $\theta_0 \ll \lambda/r_b$ and $\Delta\nu \ll \Delta\nu_b$:

Image	sub-images	t_{int}
SNAPSHOT	coherent	$vt_{\text{int}} < r_b$
AVERAGE	incoherent	$r_b < vt_{\text{int}} < r_F^2/r_b$
ENSEMBLE-AVERAGE	random	$vt_{\text{int}} = \infty$

Cornwell[2] calls these the "diffractive," "refractive," and "super-refractive" images. For the typical bandwidths and integration times used in VLBI, the average/refractive image is the most relevant.

3. NUMERICAL SIMULATIONS

Numerical simulations with one-dimensional phase screens confirm the theoretical picture presented above. (To be honest, the simulations came first, and our insight

Imaging Techniques

developed as we struggled to understand them.) Fresnel diffraction theory, rather than geometrical optics, was used, and the diffraction integral was done by means of 2^{19}-point FFT's. Realizations of the phase were based on the power spectrum

$$Q(q) = Q_0(q^2 + q_{min}^2)^{-\beta/2}\exp(-q/q_{max}), \qquad (4)$$

where q_{min} and q_{max} are reciprocally related to outer and inner scales r_{out} and r_{in}, respectively. For the "Kolmogorov" Type A spectrum and a *two*-dimensional screen, $\beta = 11/3$; in our one-dimensional analog, $\beta = 8/3$. All of our simulations corresponded to very strong scintillation, with $r_{ref}/r_F > 100$.

In addition to the conclusions already stated above, we found that

1. For Type A spectra, the average image is smooth and closely approximates the ensemble-average image.
2. A Type B spectrum ($\beta > 4$ for 2-D screens, or $\beta > 3$ for 1-D) gives deeply modulated, *fractal* average images. On theoretical grounds (see below), we expect the ensemble-average image to be very broad in this case, because of large image wander, and smooth.
3. Imposing an inner scale larger than r_F on a Kolmogorov spectrum (an example of Narayan's Type A1) leads to obvious extra structure within the image.

4. PHYSICAL OPTICS

Quantitative interpretations of the VLBI data must be based on physical optics rather than the heuristic geometric optics of Section 2. Among the relatively tractable quantities are the moments of the visibility $V(\mathbf{b}) \equiv E(0)E^*(\mathbf{b})$, where $E(\mathbf{R})$ is one polarization of the narrow-band electric field at the point $\mathbf{R}$ in the observer's plane. Equation (3) gives a relation for the mean visibility, but this tells us only about the ensemble-average image, which may differ from the observationally more relevant average image. The next interesting moment is the mean-square visibility:

$$\Gamma_4(\mathbf{b}) \equiv \langle V(\mathbf{b})V^*(\mathbf{b})\rangle = \langle E(0)E^*(\mathbf{b})E^*(0)E(\mathbf{b})\rangle = \langle F(0)F(\mathbf{b})\rangle, \qquad (5)$$

where $F(\mathbf{R}) \equiv |E(\mathbf{R})|^2$ is the flux received at $\mathbf{R}$. From this and the mean visibility (5) we can construct $\langle|\Delta V(\mathbf{b})|^2\rangle$, where $\Delta V(\mathbf{b}) \equiv V(\mathbf{b}) - \langle V(\mathbf{b})\rangle$, and thus quantify the typical departure of the measured visibility from its ensemble average. Equation (5) as written actually gives the mean-square *snapshot* visibility; a generalization is required to get at the average visibility[6].

Equation (5) allows us to apply established results for the flux correlation to our study of the visibility. In strong scintillation, $\Gamma_4(\mathbf{b})$ can be written as the sum of three parts. The first is $\langle F\rangle^2$. The second is positive and independent of baseline for $b > r_{diff}$; this apparently describes the white-noise structure in the snapshot image, because it is filtered out when we pass to the average image. The third and most interesting term depends on b and is present in the average as well as the snapshot visibility. It describes extra structure or "colored noise" in the average image that is absent from the ensemble-average image. For two-dimensional screens, the b-dependence of this term is $\propto b^{2-\beta}$ for Type A spectra ($2 < \beta < 4$) and $\propto b^{\beta-6}$ for Type B ($4 < \beta < 6$); in the former case, however, there is a small coefficient that vanishes for infinitely strong scintillation, and the total power in this term (i.e. $\int 2\pi b db \ldots$) is $\sim (r_{ref}/r_F)^{-2(4-\beta)}$. Thus this contribution to $\langle|\Delta V(\mathbf{b})|^2\rangle$ is weak for the standard "Kolmogorov" case, but since it falls only as a power of b, it can dominate for $b \gg r_{diff}$, where the ensemble-averaged visibility (3) is exponentially small. For Type B spectra, the power in this term is of order unity, implying a strongly fragmented average image.*

RN thanks T. J. Cornwell and K. R. Anantharamaiah for useful discussions. This work was supported in part by an NSF grant, AST 86-11121, a New Jersey High Technology Grant, 88-240090-2, and by a W.M. Keck Foundation Fellowship.

REFERENCES

1. P.N. Wilkinson, R.E. Spencer, and R.F. Nelson, in IAU Colloquium 129, The Impact of VLBI on Astrophysics and Geophysics (Reidel, Dordrecht, 1987)
2. T.J. Cornwell, K.R. Anantharamaiah, and R. Narayan, this workshop.
3. C. Gwinn, N. Bartel, J. Cordes, A. Wolsczan, and R. Mutel, this workshop.
4. D. Backer, this workshop.
5. A. Fey, S. Spangler, and R. Mutel, this workshop.
6. S. Spangler and R. Mutel, this workshop.
7. R. Mutel, this workshop.
8. R. Narayan and J. Goodman, Mon. Not. R. astr. Soc., submitted (1988).
9. T.J. Cornwell and P.J. Napier, in Proc. NRAO Greenbank Workshop on "Radio Astronomy in Space," (1986).
10. F. Roddier, Progress in Optics XVII, 208 (1981).
11. J. Goodman, R.W. Romani, R.D. Blandford, and R. Narayan, Mon. Not. R. astr. Soc. 229, 73 (1987).
12. R. Narayan, this workshop.
13. V.I. Tatarskii and V.U. Zavorotnyi, Progress in Optics XVII, 208 (1980).

---

* There seemed to be some confusion at this workshop about the visibility expected for Type B spectra. It has been claimed that the visibility in this case should be "gaussian". Indeed, for $\beta > 4$, $\log\langle V(\mathbf{b})\rangle$ is dominated by a term $\propto -b^2 r_{out}^{\beta-4}$. However, this term describes image *wander* on a timescale $\sim r_{out}/v$, not *broadening*. Since standard VLBI techniques recenter the image, this term is filtered out, which is just as well, since the wander diverges and $\langle V(\mathbf{b})\rangle \to 0$ as $r_{out} \to \infty$ for $\mathbf{b} \neq 0$. On the other hand, the higher moment (5) is insensitive to wander, and the r.m.s. visibility for the average image is the square root of (5) [after subtraction of the constant second term due to white noise in the snapshot image], which is not a gaussian.

Computer Modeling of Interstellar Scattering Effects on Pulsar Timing

ROGER S. FOSTER

Astronomy Department and Radio Astronomy Laboratory
University of California at Berkeley

and

JAMES M. CORDES

Astronomy Department
Cornell University

Abstract

The preliminary results of a computer simulation to study the effects of interstellar scattering on pulsar pulse time-of-arrivals are presented. A one-dimensional power-law and discrete event phase screen has been used to examine the dominant ISS terms due to dispersion measure fluctuations and refraction. The main goal of this simulation is to see how often and at how many frequencies arrival times need to be measured in order to remove the plasma propagation effects. Removal of these effects is necessary to obtain accurate estimates of the proper motion and parallactic distance.

Introduction

A computer model has been devised to study the time-of-arrival variations associated with the low wavenumber, refractive component of the ISM. This effort was motivated by the effect an electron density turbulence spectrum might have on pulsar timing residuals (e.g. Armstrong 1984; Blandford, Narayan, and Romani 1984; Cordes, Pidwerbetsky, and Lovelace 1986). A one-dimensional phase screen simulates the effects of a varying electron column density along different lines of sight. The electron density irregularity spectrum is modelled as a power spectrum

$$P_{\delta n_e}(q) = C_N^2 q^{-\alpha}, \qquad q_0 \leq q \leq q_1, \qquad (1)$$

where q is the fluctuation frequency, q_0 and q_1 are the low frequency and high frequency cut-offs respectively, C_N^2 measures the amplitude of the power spectrum, and α is the spectral index. A Kolmogorov spectrum in the one-dimensional case corresponds to $\alpha = 8/3$. The phase delay $\phi(x)$ as a function of position along the screen can be directly related to the changing electron column density as follows:

$$\phi(x) = -\lambda r_e \int_0^{\Delta z} dz' \delta n_e(x) = -8.4 \times 10^5 \lambda \Delta DM, \qquad (2)$$

where r_e is the classical electron radius, λ is the observing wavelength in cm, $\delta n_e(x)$ is the amplitude of the electron density fluctuations, ΔDM is the dispersion measure fluctuation (in pc cm^{-3}) and Δz is the screen thickness. The refractive angle produced by the screen is proportional to the gradient of the phase across the screen:

$$\theta_r(x) = k^{-1} \frac{\partial \phi(x)}{\partial x}, \qquad (3)$$

where k is the wavenumber.

The phase screen $\phi(x)$ produces three major delay terms that can be observed in pulsar timing analysis. The dispersion measure delay is proportional to the change in the phase divided by the frequency, $\sim \delta\phi(x)/\nu$. The geometrical time delay, $\sim D\theta_r^2/2c$ depends on the increased path-length a wave travels after refractive bending. Finally, the angle-of-arrival fluctuations introduce uncertainties in estimating the time of arrival at the solar system barycenter, this manifests itself as an additional refractive delay term. This ISS barycentric correction error term for a source in the ecliptic plane is $\sim (AU\theta_r/c)cos(\Omega t)$, where Ω is the orbital frequency of the earth around the sun. In the simple single cloud model these three delays scale as λ^2, λ^4, and λ^2 respectively.

Other frequency dependent delays that contribute at the sub-microsecond level, including variability of the diffraction smearing time, imperfect polarization calibration, and pulse phase jitter. These additional delays have been explicitly neglected in this simulation, but they are effectively included in the white noise background. Table I lists the various delay terms, the wavelength dependence of each term, and the amplitude of the term scaled for PSR 1937+21 at an observing frequency of 1 GHz.

Various phase screens have been used in the simulations, but most of the work was done using a standard Kolmogorov power-law screen. Several extreme scattering events (ESE's) were simulated and are shown to be easily observable in the timing residuals of our model pulsar.

Model

The simulation model is based on a spherical wave radiating from a pulsar at distance D. A one-dimensional phase screen is generated and propagated perpendicular to the observer at a velocity V_{scr} and distance $D/2$. The screen has explicitly mapped structure covering scales from $\sim 10^{12}$ cm to $\sim 10^{16}$ cm. Length scales between 10^9 cm to 10^{12} cm are reached by scaling laws. Each point on the phase screen is mapped to a unique point on the ground using a ray-optics code. Every point on the ground then has an accumulated intensity and average angle-of-arrival. The 10^{16} cm outer scale is arbitrary, and is simply an effect of the size of the computed phase screen.

The resulting data scale with frequency according to the observational input parameters, the turbulent spectrum and the assumed scintillation bandwidth. The arrival times of the wavefronts are adjusted to topocentric arrival times from barycentric arrival times assuming a circular earth orbit. The data are fit for standard pulsar spin parameters $(P, \dot{P}, \ddot{P})$, proper motion and distance, then residuals are output for further analysis. Table II displays the assumed parameters for the PSR 1937+21 system used in the simulation (Rawley et al. 1987, 1988). The timing accuracy was (optimistically) assumed to be ~ 100 nanoseconds for most runs.

Results

Assuming a Kolmogorov power law spectrum for the electron turbulence and an observing frequency of 1 GHz, a four-year simulation for one particular phase screen produces ISS barycentric correction errors of ~ 0.5 μs, geometrical delays of ~ 0.3 μs, and dispersion measure delays of about 4 μs. These terms plus the final fits are shown in figures 1 for 1 GHz and figure 2 for 500 MHz. Proper motion is determined to better than 50 percent of the true value, while the distance estimate has no statistical significance. For $\alpha = 8/3$ at 500 MHz the timing observations are degraded. The ISS barycentric correction error term broadens to ± 1 μs, while the peak of the geometrical delay increases to more than 4 μs, and the dispersion measure delay increases to ~ 12 μs peak-to-peak. At both frequencies the final residuals show the underlying dispersion measure delays.

Lowering the spectral index of the power-law to 7/3 improves the distance estimate dramatically to within 20 percent of the true value for the 1 GHz simulation. Dispersion measure and geometrical delay terms cannot be separated by single frequency observations. Dual frequency observations fitting for both λ^2 and λ^4 terms could help separate the various frequency dependent terms. The ISS barycentric correction term is not easily separable from the dispersion measure fluctuations because they both scale as λ^2 but have different time dependences. At observing frequencies near 1 GHz, dispersion measure and ISS barycentric correction errors dominate for a Kolmogorov spectrum and four years of simulated data. Near 500 MHz the geometrical term becomes more important.

An extreme scattering event was modeled by the simulation with an effective ΔDM of 0.02 pc cm^{-3}. The event is obvious as a discontinuity in the timing residuals. Figure 3 shows the dramatic effect a caustic event would have on pulsar timing data. Clearly the Arecibo timing data from Rawley et al. (1988) does not show any such event.

Conclusions

Fitting for a λ^2 frequency dependence between multi-frequency data should be productive in removing the effects of dispersion measure variation. Separating out the λ^2 dependence of the ISS barycentric correction term and the λ^4 geometrical delay may be more subtle. Assuming a particular form for the turbulence spectrum will allow for a better frequency dependent fit.

High precision timing data require very frequent regular sampling over intervals much less than the refraction time scale. A good distance estimate from the parallax term requires sampling at intervals of less than one month. Angle-of-arrival and dispersion measure fluctuations may be evident on time scales of days. Comparison of observational and simulated data for particular phase screens and power spectra could help remove some of the frequency dependent timing errors.

Long-term high precision timing of multiple pulsars along different lines of sight can place constraints on the likelihood of observing ESE's of the kind reported by Fiedler et al. (1987). Well sampled data also place constraints on the electron density turbulence spectrum that exists in the interstellar medium.

The authors gratefully acknowledge helpful discussions with R. Romani and D. Backer. Particular mention is given to D. Backer for his comments on an earlier draft of this paper and his suggestion to consider the ISS barycentric correction error effect on pulsar timing.

References

Armstrong, J. W. 1984, *Nature*, **307**, 527.

Blandford, R. D., Narayan, R., and Romani, R. 1984, *J. Astr. Ap.*, **5**, 369.

Cordes, J. M., Pidwerbetsky, A., and Lovelace, R. V. E. 1986 *Ap. J.*, **310**, 737.

Fiedler, R., Dennison, B., Johnston, K., Hewish, A. 1987 *Nature*, **326**, 675.

Rawley, L. A., Taylor, J. H., Davis, M. M., and Allan, D. W. 1987, *Science*, **238**, 761.

Rawley, L. A., Taylor, J. H., Davis, M. M. 1988, *Ap. J.*, **326**, 947.

TABLE I

TIME OF ARRIVAL VARIATIONS

Term	Single cloud	Kolmogorov spectrum	PSR 1937 @ 1 GHz (τ in months)
DM variation	λ^2	λ^2	$0.4\,\tau^{5/6}\ \mu s$ ($\tau_{sat} \approx 10^4$ yr)
Geometric time delay	λ^4	$\lambda^{49/15}$	$0.1\,\tau\ \mu s$ ($\tau_{sat} \approx 2.3$ months)
ISS 'barcentric error'	λ^2	$\lambda^{49/30}$	$0.2\,\tau\cos(\Omega t)\ \mu s$ ($\tau_{sat} \approx 1.9$ months)

TABLE II

ASSUMED PARAMETERS FOR PSR 1937+21

Term		Assumed value
Scintillation Bandwidth	$\Delta\nu$	3×10^5 Hz
Period	P	1.557806 ms
Period Derivative	$\dot{P}$	$1.051 \times 10^{-19}\,\text{s}\,\text{s}^{-1}$
Position	$R.A.$	$19^h\ 37^m\ 28^s$
	$DEC.$	$21°\ 28'\ 01''$
Proper motion	μ_α	$-2.59 \times 10^{-4}\ ''/\text{yr}$
	μ_δ	$-4.36 \times 10^{-4}\ ''/\text{yr}$
Distance	D	5 Kpc
Observing frequency	ν	1 GHz
Slope of Turbulence Spectrum	α	2.66

Imaging Techniques

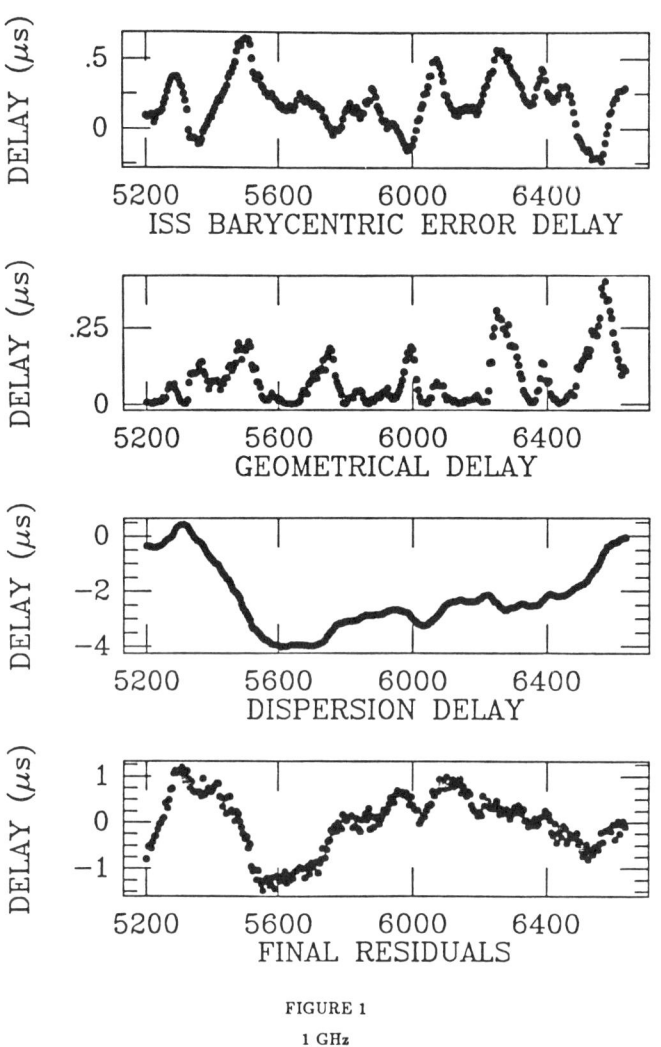

FIGURE 1

1 GHz

Simulated timing residuals at 1 GHz as a function of observation date. Panels show the ISS barycentric correction term, geometrical delay, dispersion delay, and post-fit residuals.

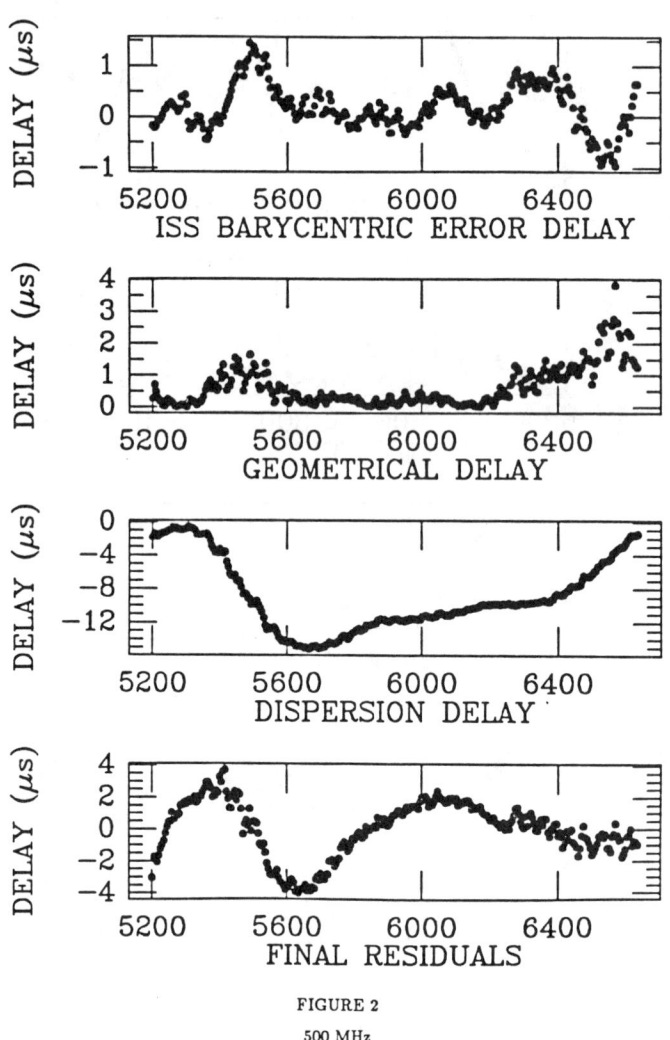

FIGURE 2
500 MHz

Simulated timing residuals at 500 MHz as a function of observation date. Panels show the ISS barycentric correction term, geometrical delay, dispersion delay, and post-fit residuals.

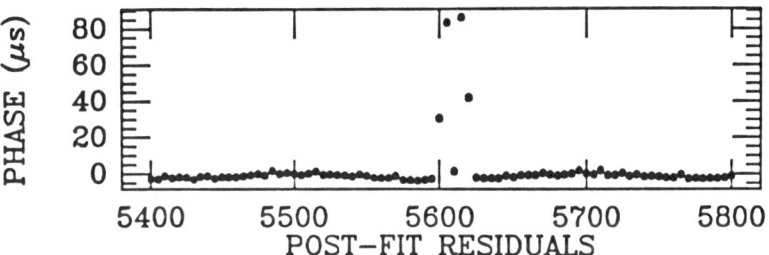

FIGURE 3

Post-fit timing residuals for an Extreme Scattering Event (ESE) due to an effective electron column density of $\Delta DM \sim 0.02$ pc cm^{-3}.

INTERSTELLAR INTERFEROMETRY

J. M. Cordes and A. Wolszczan
National Astronomy and Ionosphere Center

ABSTRACT

Interstellar scintillations may be used to resolve pulsar magnetospheres, especially during multiple imaging events caused by discrete lenses in the interstellar plasma. Multiple paths induce fringes in dynamic spectra analogous to those of an interferometer. The effective baseline length $\sim D_\ell \theta_{split} \sim 1$ AU for a lens distance $D_\ell \sim 1$ kpc and splitting angle $\theta_{split} \sim 1$ mas, yielding resolution <0.1 $microarc$ sec. We derive a general expression relating the fringe phase to source position for an arbitrary two dimensional lens. The fringe phases derived from standard models of the pulsar magnetosphere are much smaller than those measured. We discuss possible modifications of the standard pulsar model and we also speculate on applications of the technique to active galactic nuclei.

RESOLVING PULSAR MAGNETOSPHERES

Diffractive interstellar scintillations (DISS) are quenched if the source angular diameter $\gg \theta_c \approx \lambda/2\pi\theta_d D \approx 0.07[\nu_{GHz}\theta_d(m.a.s.)D_{kpc}]^{-1}$ μas where θ_d is the size of the scattering disk and D is the source distance. For comparison, the angular diameter of a pulsar light cylinder is $\theta_{LC} = 0.33 P D_{kpc}^{-1}$ μas. Lovelace[1] first pointed out that DISS may be used to resolve pulsar magnetospheres, an idea promulgated by Lyne[2], and used[3,4] to produce crude limits on the size of the emission region. The first strong limits[5] indicated spatial separations of emission regions $< 10^3$ km in two pulsars.

DISS is often characterized by random structure in dynamic spectra [defined as intensity as a function of frequency and time, $I(\nu,t)$]. Multiple imaging events induce periodic fringes in $I(\nu,t)$ which are more sensitive than pure DISS for resolving magnetospheres[6] when the fringe spacing is less than the characteristic diffraction bandwidth or time scale. Recently[7], the method of ref 6 was applied to the nearby pulsar PSR1237+25. Surprisingly, measured fringe phases were much larger than those expected for a standard model of the pulsar magnetosphere. Here, we analyze multiple imaging events using two dimensional lenses, derive several closure relations, discuss the PSR1237+25 results, and discuss how future observations will teach us about both pulsar magnetospheres and interstellar plasma lenses.

MULTIPLE IMAGING EVENTS

Figure 1 shows dynamic spectra[7] during a multiple imaging event. The frequency of occurrence of these events is poorly known because they require copious amounts of telescope time to monitor and because both low and high Q oscillations occur. Cases devoid of oscillations sometimes show evidence of multiple images through the appearance of nonrandom criss-cross patterns in dynamic spectra[8,9].

Two scintillation surveys made at the Arecibo Observatory provide the best constraints, so far. In Survey I[8], 31 pulsars were observed in each of two weeks of 1980 and 1981. A visual assessment suggests that (1) 4 of 31 pulsars show obvious periodicities; (2) 10 of 31 show nonrandom scintillations ('nonrandom' as described above, does not include simple frequency drifts in the frequency-time plane of otherwise random fea-

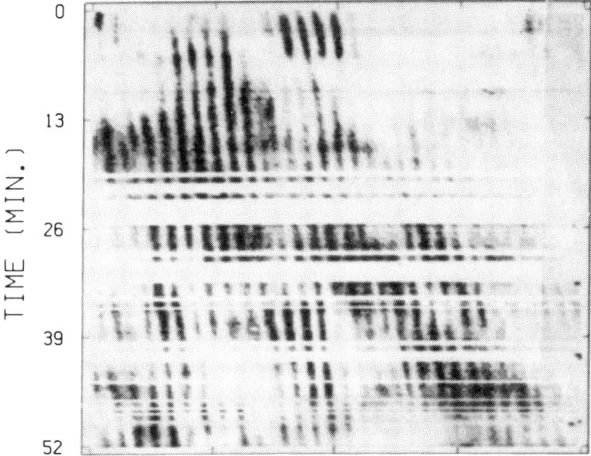

Figure 1 Dynamic spectra across a 10 MHz bandwidth at 430 MHz for the pulsar 1237+25 (after Wolszczan and Cordes 1987). Horizontal stripes correspond to excised interference.

tures); (3) 17 of 31 show essentially random scintillations. In Survey II[9] observations of 9 pulsars were made for 3 consecutive days in each of 12 months in 1984-85. Daily observations were made in June 1985 in collaboration with B. Rickett. The overall results imply that 5 of 9 pulsars show periodic fringes or criss cross patterns at least some of the time.

FRINGE PHASE FOR TWO DIMENSIONAL LENSES

Let D and D_ℓ be the earth and lens distances from the pulsar. For a source at transverse position $\vec{x}_s$ (from a line connecting the earth with the neutron star) a pair of images yields a total dynamic spectrum $I = I_1 + I_2 + 2(I_1 I_2)^{1/2} \cos[\Phi_{12}(\vec{x}_s) + \phi_{12}]$, where all quantities are functions of ν, t. In the following we concentrate on the geometrically determined fringe phase Φ_{12} and ignore the amplitude modulation $I_1 I_2$ and phase modulation ϕ_{12} caused by diffraction. For a cold plasma, $\Phi_{12} \propto \nu^{-3}$ and varies with time by virtue of motions of source, lens, and observer. We assume that source motion dominates other motions. The fringe phase may be expanded in frequency, time, and transverse source position:

$$\Phi_{12} = \Phi_{12}(\nu_0, t_0, \vec{x}_s) + \nabla_{\perp x_s} \Phi_{12} \cdot \delta \vec{x}_s + \frac{2\pi t}{P_t} + \frac{2\pi \nu}{P_\nu}. \qquad (1)$$

The fringe periods P_ν and P_t are

$$P_\nu \equiv \frac{2\pi}{|\partial_\nu \Phi_{12}|} = \frac{2\pi \nu}{3\Phi_{12}} = \frac{2c}{3D_\ell (1 - D_\ell/D)|\theta_2^2 - \theta_1^2|} \qquad P_t \equiv \frac{2\pi}{|\nabla_{\perp x_s} \Phi_{12} \cdot \vec{V}_\perp|}. \qquad (2)$$

The temporal period P_t depends on the proper motion velocity $\vec{V}_\perp$ but not the radial motion. The periods scale as $P_t \propto \nu$ and $P_\nu \propto \nu^4$. At large ν, P_ν becomes larger than the receiver bandwidth and the fringe pattern is essentially a temporal variation.

Typical image splitting angles are $\theta_{split} \equiv \sqrt{\theta_2^2 - \theta_1^2} = 0.5 f(D_\ell/D) [DP_\nu]^{-1/2}$ mas corresponding to a column density (expressed as a dispersion measure change) $\delta DM \equiv \int_0^D dz\, n_e \sim 10^{14} f(D_\ell/D) \nu^2 \ell_\perp \theta_{split} [DP_\nu]^{-1/2}$ cm^{-2} where $f(x) \equiv [x(1-x)]^{-1/2}$, D is in kpc, ν in GHz, P_ν in MHz, θ_{split} in mas, and the transverse length of the lens, $\ell_\perp$, is in AU.

We may express the source displacement $\delta \vec{x}_s$, source velocity $\vec{V}_\perp$, and lens gradient $\nabla_{\perp x_s} \Phi_{12}$ in terms of their magnitudes and position angles χ_{x_s}, χ_V, and χ_∇, respectively. Then the phase shift between two sources separated by $\delta \vec{x}_s$ is

$$\delta \Phi_{12} = \frac{2\pi \delta x_s \cos(\chi_{x_s} - \chi_\nabla)}{V_\perp P_t |\cos(\chi_V - \chi_\nabla)|}. \tag{3}$$

The position angle of the proper motion, χ_V, is time invariant, whereas the source position angle, χ_{x_s}, varies with pulse phase and χ_∇, the position angle of the lens gradient, is epoch dependent, varying slowly on 'refractive' time scales of order days to months (e.g. Rickett, this volume). By virtue of these different variations, it is possible to learn about both the pulsar magnetosphere and the interstellar medium by measuring the fringe phase across pulse phase at many epochs. The shape of the fringe curve depends on the difference in position angles, $\chi_{x_s} - \chi_\nabla$. Therefore, observations at different epochs are likely to show: (a) Changes in both amplitude and shape of the fringe phase curve vs. pulse phase; (b) Whether χ_∇ is isotropic or whether it has a preferred direction. The latter case would signify that the ISM is anisotropic on refractive length scales ~ 1 AU.

Phase screens are expected to yield *odd* numbers of images. When N images occur, there are $N_p \equiv N(N-1)/2$ unique pairs of images. The pair composed of the ith and j^{th} images is characterized by a fringe pattern with periods $P_{\nu ij}, P_{t ij}$. Since the periods are determined by differences of *pointwise* calculated lens parameters, various closure relations are expected, analogous to those associated with interferometer systems. Consider the ith, j^{th}, and k^{th} images. The closure phase is $\Phi_{ijk} = \Phi_{ij} + \Phi_{jk} + \Phi_{ki} = 0$. Therefore the fringe phase difference closes: $\delta \Phi_{ijk} = \delta \Phi_{ij} + \delta \Phi_{jk} + \delta \Phi_{ki} = 0$. Since the derivatives of Φ_{ijk} vanish, reciprocals of the periods $P_{t_{ij}}$ and $P_{\nu_{ij}}$ sum to zero around closed loops (when weighted by the *signs* of $\partial_\nu \Phi_{ij}$ and $\nabla_{\perp x_s} \Phi_{ij}$, respectively). The closure relations imply that if any two periods can be identified in the data, then identification of the third may be aided using a prediction from the closure relation.

FRINGE PHASE FOR STANDARD PULSAR MODELS

The standard model for radio emission from pulsar magnetospheres entails relativistic flow near the axis of an approximately dipolar magnetic field. The essential features of the model include: (1) Strong relativistic beaming with particle Lorentz factors $\gamma \gg 1$; (2) Constant altitude of emission r_{em} for a given radio frequency with r_{em} much less than the light cylinder radius, $r_{LC} \equiv c/\Omega$; (3) Determination of the plane of polarization by a frozen in, roughly dipolar magnetic field; and (4) No refraction of radio waves within the magnetosphere.

The standard model accounts for many pulsar properties, at least in a broad brush way. Particularly compelling is the congruence of predicted and measured polarization curves against pulse phase, as in the original proposal by Radhakrishnan and Cooke[10]. The standard model may be used to predict the fringe phase $\delta \Phi_{12}$ as a function of pulse phase η; fringe phases vary because emission region locations on the plane of the sky vary with pulse phase[5]. The position angle of the emission region is the same as the polarization position angle,

$$\chi_{x_e} = \chi_\Omega + \tan^{-1}\left[\frac{\sin\alpha\sin\eta}{\sin(\beta-\alpha)+(1-\cos\eta)\cos\beta\sin\alpha}\right], \qquad (4)$$

where χ_Ω is the position angle of the spin axis; α and β are the rotational colatitudes of the dipole axis and line of sight, respectively.

Figure 2 shows intensity, polarization position angle, and predicted fringe phase plotted against pulse phase for $\alpha = 60°$, $\beta = 65°$, and $r_{em}/r_{LC} = 0.1$ for a pulsar with spin period $P = 1$ sec and proper motion velocity 100 km s^{-1}. The assumed fringe period is $P_t = 1000$ sec and two values of the lens position angle have been used (filled and open circles in top panel). As is obvious, the fringe phase varies by only a few degrees over the pulse duration for these parameters.

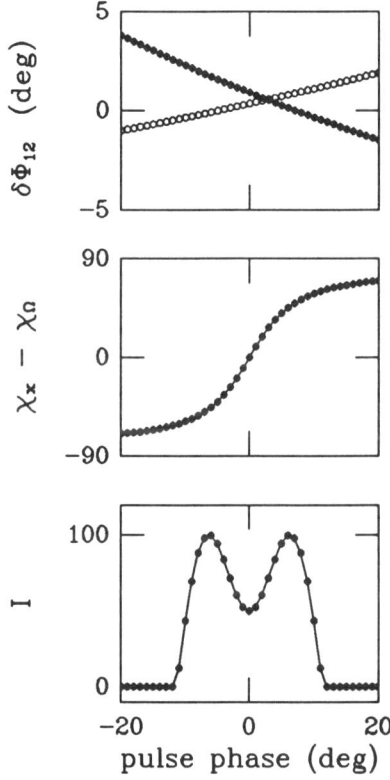

Figure 2 Predicted waveforms.

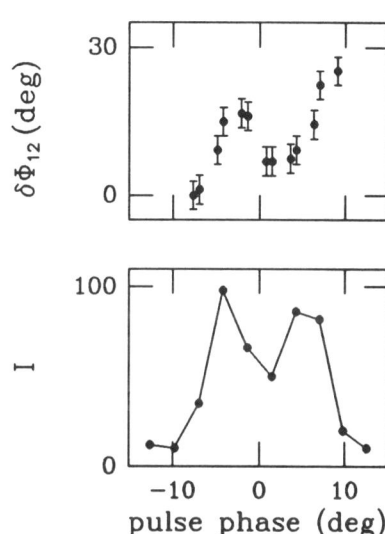

Figure 3 Waveform and fringe phase curve for PSR1237+25.

Figure 3 shows the pulse shape and fringe phase curve reported in ref 7 for the pulsar 1237+25 ($P = 1.3$ s, $V_\perp = 178$ km s^{-1}), when $P_t \approx 1300$ s. The fringe phases imply transverse separations ~few $\times 10^8$ cm $\ll r_{LC} = 10^{9.8}$ cm. However, when compared to the standard model, it is clear that the measured fringe phase variation is about 10 times larger than that in Figure 2. The small differences in P and $V_\perp$ do not account for the differences. We conclude that the large phases require: (1) that emission altitudes $\approx$ the light cylinder radius; or (2) that the field line structure has smaller radii of curvature than predicted for a dipolar field. For case (1), emission altitudes may actually mean altitudes where free space propagation ensues if radiation is significantly refracted in the magnetosphere[11]. Large emission or decoupling altitudes result in significant aberration of beamed radiation, some of which may be canceled by sweep back of magnetic field lines. Note that the unknown lens distance has no influence on these conclusions, since it is contained in P_t, which is a measureable quantity.

DISCUSSION

The remarkable conclusion that we have come to is that interstellar scintillation observations may be used to resolve pulsar magnetospheres. So far, results suggest that emission regions radiate from locations that vary by a few $\times 10^8$ cm over the observed pulse components. Another implication is that relativistic beaming angles must be smaller than $2\delta\eta/3$, where $\delta\eta = 2°.8$ is the pulse phase resolution used in ref 7; this yields $\gamma > 30$.

Future work should aim at (1) determining fringe phase curves for a statistically meaningful sample of pulsars; (2) multiepoch measurements to determine the stability of fringe phase curves, thus providing information on the shapes and orientations of plasma lenses in the interstellar medium; (3) further modeling of pulsar magnetospheres, including aberration and distortions of dipolar fields; (4) a search for fringes from compact extragalactic sources, particularly during strong refraction events[12] when multiple images are likely to form; any fringes seen would imply brightness temperatures well in excess of 10^{12} K, and the likely presence of coherent emission.

This work was supported by the National Astronomy and Ionosphere Center at Cornell, which operates the Arecibo Observatory under contract to the National Science Foundation. This work was also supported by NSF grant 85-20530 to Cornell University.

REFERENCES

1. Lovelace, R.V.E. 1970, *Ph.D. Thesis*, Cornell University.
2. Lyne, A. G. 1974, talk at Stanford Pulsar Symposium, Palo Alto.
3. Backer, D. C. 1975; *Ast. Ap.*, **43**, 395.
4. Manchester, R.N. and Lyne, A.G. 1984; private communication.
5. Cordes, J. M., Weisberg, J., and Boriakoff, V. 1983, *Ap. J.*, **268**, 370.
6. Cordes, J.M., Pidwerbetsky, A., and Lovelace, R. 1986; *Ap. J.*, **310**, 737.
7. Wolszczan, A. and Cordes, J.M. 1987; *Ap. J. (Letts)*, **320**, L35.
8. Cordes, J. M., Weisberg, J., and Boriakoff, V. 1985, *Ap. J.*, **288**, 221.
9. Cordes, J.M. and Wolszczan, A. 1986; *Ap. J. (Letts)*, **307**, L27.
10. Radhakrishnan, V. and Cooke, D. J. 1969; *Ap. Letters*, **3**, 225.
11. Barnard, J.J. and Arons, J. 1986, *Ap. J.*, **302**, 138.
12. Fiedler, R. L., Dennison, B., Johnston, K.J., and Hewish, A. *Nature*, **326**, 675.

INTERSTELLAR SCINTILLATIONS OF BINARY PULSARS

R. J. DEWEY, J. M. CORDES A. WOLSZCZAN, AND J. M. WEISBERG

Abstract

We have measured interstellar scintillations of two binary pulsars (1855+09 and 1913+16) to determine their transverse velocities at numerous points in their orbits. Combining transverse velocities with radial velocities derived from Doppler shifts, one can (in principle) determine a binary system's proper motion and orbital inclination, and the bulk velocity of the scattering material along the line of sight. With our present observations we cannot unambiguously determine these quantities but we are able to place interesting limits on some of them. In particular, we find that the orbital inclination of 1855+09 is large ($> 70°$), a finding which strongly constrains the mass of the pulsar's companion. The bulk velocity of scattering material seems to be less than 25 $\mathrm{km\,s^{-1}}$.

Introduction

It is possible to measure a pulsar's transverse speed using the time variation of its interstellar scintillation pattern. This technique has been used to measure the speeds of a large number of isolated pulsars (Cordes 1986) and produces results in good agreement with interferometric observations (Lyne and Smith 1982). Since these measurements can be made with less than an hour of data, they are ideally suited to following the changing speed of a binary pulsar as it moves in its orbit (Lyne 1984), thereby allowing determinations of the orbital inclination and the transverse space velocity ot the pulsar.

Most theories relate the magnitude of a pulsar's velocity directly to the circumstances in which it was formed – the orbital speed of the progenitor star when it exploded, the amount of mass ejected by the supernova, and the degree of symmetry of that mass ejection. Knowledge of the orbital inclination is useful since, with other known parameters, it determines the system's separation and the mass ratio of the two objects, quantities which in turn constrain the possible evolutionary history of the system.

We measured the scintillation velocities of two binary pulsars, 1913+16 ($P_{orb} = 7.75\,\mathrm{h}$) and 1855+09 ($P_{orb} = 12.33\,\mathrm{d}$), at numerous points in their orbits. The evolutionary histories of these pulsars probably differed significantly and their space velocities are expected to reflect this. Standard scenarios predict a velocity on the order of hundreds of $\mathrm{km\,s^{-1}}$ for 1913+16, and a velocity at least an order of magnitude lower for 1855+09. Our observations are consistent with these predictions though they do not place interesting limits on possible evolutionary scenarios. For 1855+09 we are able to constrain the system's orbital inclination (for 1913+16 it was already known) and in doing so limit the possible mass range of the companion. Our results are in good agreement with theoretical work on the nature of the companion (Savonije 1987).

Scintillation Speed

The pulsar's scintillation speed v_{iss} is inversely proportional to the characteristic timescale τ of the scintillation pattern and is the transverse component of an effective vector velocity that is the weighted vector sum of the pulsar's velocity, $\mathbf{v}_{psr}$, the earth's velocity, $\mathbf{v}_\oplus$, and the bulk motion of the scattering material, $\mathbf{v}_{ism}$:

$$\mathbf{v}_{eff} = (1-f)\mathbf{v}_{psr} + f\mathbf{v}_\oplus - \mathbf{v}_{ism},$$

where f is the ratio of the pulsar-scattering screen distance. For binary pulsars $\mathbf{v}_{psr}$ is the sum of a constant center of mass velocity (the transverse components of which correspond to the pulsar's proper motion), and a time varying orbital velocity, so $\mathbf{v}_{eff}$ can be written:

$$\mathbf{v}_{eff} = (1-f)\mathbf{v}_{orb}(t) + f\mathbf{v}_\oplus + (1-f)\mathbf{v}_{pm} - \mathbf{v}_{ism}, \tag{1}$$

On time scales much less than a year any time varying component of $\mathbf{v}_{eff}$ (and hence in v_{iss}) is related directly to the pulsar's orbital motion.

Of the quantities on the right-hand-side of equation (1) only $\mathbf{v}_\oplus$ and the radial component of $\mathbf{v}_{orb}$ are known unabiguously. In addition, the constant that relates v_{iss} to τ is uncertain (possibly by as much as a factor of two) due to uncertainty in the pulsar distance and the statistics of the scattering material. In principle, however, all these unknowns can be determined empirically with measurements at enough points in the orbit.

Binary Motion

Pulse timing observations determine the radial component of the pulsar's orbital motion, and thus the system's eccentricity, the epoch of the ascending node, the epoch of periastron and the system's mass function. In the case of 1913+16 general relativistic effects have also determined the inclination angle, though for most systems it is not possible to do this. Timing measurements do not determine the orientation of the line of nodes in the plane of the sky. The timing parameters of 1855+09 are $P = 5.36$ ms, $e = 0.000021$, $\omega_0 = 279°$, $T_0 = 2446421.0$ JED (Rawley 1986) while the parameters for 1913+16 are $P = 59.02$ ms, $e = 0.617$, $\omega_0 = 178.86°$, $T_0 = 2442321.433$ JED, $\dot\omega = 4.23°\,\mathrm{y}^{-1}$, and $i \approx 50°$ (Weisberg and Taylor 1984).

Since the inclination of the 1913+16 system is known, the magnitude of the transverse component of $\mathbf{v}_{orb}$ can be inferred and is shown as a function of orbital phase in Figure 1a. As Figure 1b shows, for the 1855+09 system, the transverse component of $\mathbf{v}_{orb}$ is a function of the assumed inclination angle. The pulsar's total transverse velocity depends also on the magnitude of the proper motion v_{pm} and its angle θ_{pm} to the line of nodes.

Observations and Analysis

The observations were made at Arecibo Observatory, most of them in 1987 October using the 2048 channel autocorrelator to obtain dynamic spectra (intensity vs. time and frequency). For 1913+16 we included some observations made at Arecibo in 1983 July using the older 1008 channel autocorrelator.

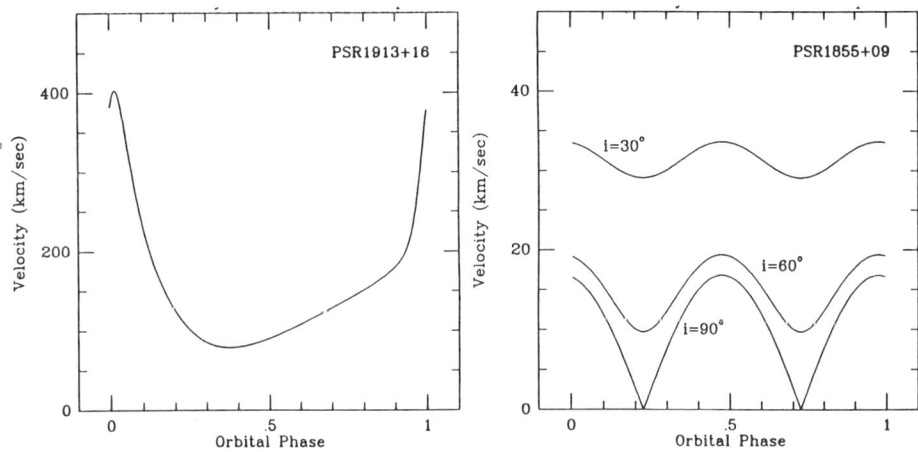

Figure 1 Transverse orbital velocity (in $km\ s^{-1}$) plotted against orbital phase: (a) PSR1913+16. (b) PSR1855+09 for different assumed inclinations.

To obtain velocities from the dynamic spectra we used the techniques described by Cordes (1986), fitting an elliptical gaussian function to the 2-D autocorrelation function of each set of dynamic spectra.

We observed 1913+16 at frequencies near 1400 MHz. About half of the 1987 observations used a total bandwidth of 40 MHz with a frequency resolution of 78 kHz, and a time resolution of 10 s; for the remainder we halved the bandwidth, doubling the frequency resolution. The 1983 observations used a 5 MHz bandwidth, with 79 kHz frequency resolution and 29 s time resolution. Fitting to the correlation functions we obtained characteristic time scales $\tau \sim 60$ s and characteristic frequency scales $\Delta\nu \sim 45$ kHz. Figure 2a shows the resulting scintillation velocities. We observed 1855+09 at 430 MHz using a total bandwidth of 10 MHz, with a frequency resolution of 156 kHz and a time resolution of 20 s. Fitting to the correlation functions we obtained characteristic time scales $\tau \sim 500$ s and characteristic frequency scales $\Delta\nu \sim 90$ kHz. The resulting velocity measurements are plotted in Figure 2b as a function of orbital phase.

Constraining the Unknowns

In modeling the expected values of v_{iss} for PSR1855+09 there are eight unknown parameters: the effective location of the scattering screen, f, the orbital inclination, i, the magnitude and direction of the system's proper motion, v_{pm}, θ_{pm}, the magnitude and direction of the bulk motion of the scattering, v_{ism}, θ_{ism}, the orientation of the binary orbit on the sky, θ_{bin}, and the overall scale factor, A, relating v_{iss} and to the scintillation time scale τ. For 1913+16 there are only seven unknowns since i is (approximately) known.

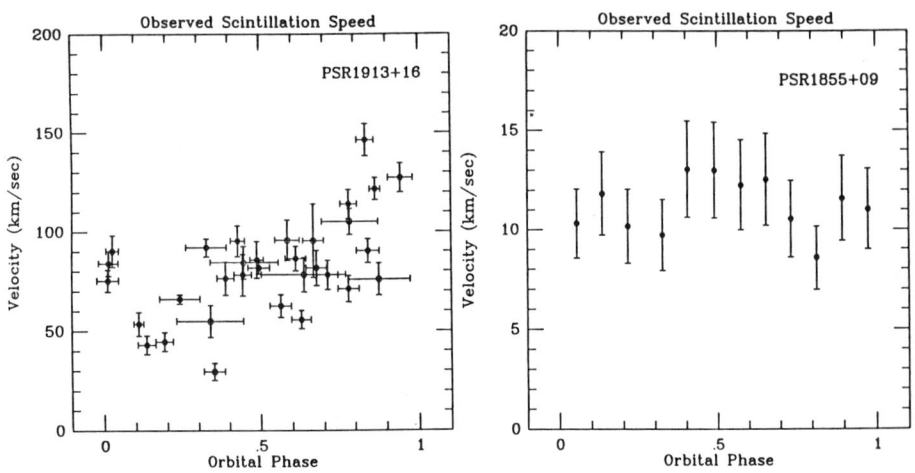

Figure 2 Scintillation speeds (in $km\ s^{-1}$) plotted against orbital phase: (a) 1913+16; (b) 1855+09.

We compared the values of v_{iss} resulting from various combinations of these parameters to the observed velocities. For both pulsars we tried four values of f from 0.2 to 0.8, three values of v_{ism} from 0 to $30\,\mathrm{km\,s^{-1}}$, five uniformly spaced angles for θ_{ism} and θ_{bin}, and scale factors, A, of 0.5, 1.0 and 2.0. For 1855+09 we tested ten values of i from 1 to 90°, eleven values of v_{pm} from 0 to $100\,\mathrm{km\,s^{-1}}$ and ten values of θ_{pm}. For 1913+16 we tried eleven values of v_{pm} from 0 to $700\,\mathrm{km\,s^{-1}}$ and ten values of θ_{pm}.

We did not find an unambiguous fit for either pulsar, but could elimate significant volumes of phase space. For 1855+09, where the earth contributes significantly to v_{iss}, observations at other times of year should help clarify the situation. On a longer time scale, the precession of the orbit of 1913+16 might help in a similar way.

Results

1855+09: The most highly constrained parameter is the orbital inclination where we find $i > 70°$, with $i \approx 90°$ quite likely. This, in conjunction with the known mass function

$$f(m_{psr}, m_{comp}) = \frac{(m_{comp} \sin i)^3}{(m_{psr} + m_{comp})^2} = 0.0052\,\mathrm{M_\odot}$$

places strong limits on the companion mass. If we assume $m_{psr} = 1.4\,\mathrm{M_\odot}$ we find $0.24\,\mathrm{M_\odot} \leq m_{comp} \leq 0.26\,\mathrm{M_\odot}$; smaller pulsar masses yield smaller companion masses. Our results agree with the theoretical work of Savonije (1987) who, using evolutionary considerations, placed an upper limit on the companion mass of $m_{comp} \leq 0.22\,\mathrm{M_\odot}$, and concluded that $i \approx 90$ and $m_{psr} \leq 1.2\,\mathrm{M_\odot}$.

The other parameters are less well determined: $f \gtrsim 0.6$ and $A > 1.0$ appear highly probable, and $v_{pm} \lesssim 80$ seems likely. However, with large values of f and A, higher values of v_{pm} are allowed.

If we fix a number of the parameters at reasonable values we can place interesting limits on the bulk motion of the scattering material. Setting $f = 0.7$, $A = 1.5$, $i = 80°$ and $v_{pm} \leq 10\,\mathrm{km\,s^{-1}}$ (likely on evolutionary grounds) and considering values of v_{ism} as large as $100\,\mathrm{km\,s^{-1}}$ we find $v_{ism} \lesssim 25\,\mathrm{km\,s^{-1}}$.

1913+16: None of the parameters of 1913+16 are very tightly constrained, though again $f \gtrsim 0.6$ and $A > 1$ appear likely. Assuming $A > 1$, a lower limit of $50\,\mathrm{km\,s^{-1}}$ can be placed on v_{pm}. The upper limit on v_{pm} is less certain, though for $A \lesssim 2$, $v_{pm} \lesssim 700\,\mathrm{km\,s^{-1}}$ appears reasonable. This is less stringent than the upper limit provided by timing data, $v_{pm} \lesssim 300\,\mathrm{km\,s^{-1}}$ (Taylor, 1988). These limits are consistent with standard evolutionary scenarios which suggest $v_{pm} \approx 170\,\mathrm{km\,s^{-1}}$ if the most recent supernova was symmetric (Cordes and Wasserman 1984). If we consider only solutions with $v_{pm} < 300\,\mathrm{km\,s^{-1}}$ we can limit the direction of the proper motion (its angle to the orbit's line of nodes) to $140° \lesssim \theta_{pm} \lesssim 250°$.

This work was supported by the National Science Foundation through Grant AST-8520530 to Cornell University, by an Ernest F. Fullham Award, and by the Alfred P. Sloan Foundation. The Arecibo Observatory is part of the National Astronomy and Ionosphere Center which is operated by Cornell University under contract with the National Science Foundation.

References

Cordes, J. M. 1986, *Astrophys. J.* **311**, 183.

Cordes, J. M. and Wasserman, I. 1984, *Astrophys. J.* **279**, 798.

Lyne, A. G. and Smith, F. G. 1982, *Nature* **298**, 825.

Lyne, A. G. 1984, *Nature* **310**, 300.

Rawley, L. A., 1986, Ph.D thesis, Princeton Univ.

Savonije, G. J. 1987, *Nature* **325**, 325.

Taylor, J. H., 1988, private communication.

Weisberg J. M. and Taylor, J. H., 1984, in *Birth and Evolution of Neutron Stars: Issues Raised by Millisecond Pulsars in Birth and Evolution of Neutron* ed. S. P. Reynolds and D. R. Stinebring, *(NRAO Workshop prodceedings) p.317.*

LOW FREQUENCY VLBI

Dayton L. Jones
Jet Propulsion Laboratory, California Institute of Technology

ABSTRACT

There are many important astrophysical questions which could be answered with observations at frequencies of a few MHz and angular resolution of an arcminute or better. For example, strong interstellar and interplanetary scattering of electromagnetic radiation could be studied along many lines of sight, and the sizes and shapes of scattering disks could be directly measured as a function of time and frequency. Such data can be obtained with an interferometer array composed of several small satellites in Earth orbit. The individual satellites for a low frequency array could be very small and inexpensive.

INTRODUCTION

The possibility of opening up a major new window in the electromagnetic spectrum for astronomical investigations is very exciting, and appears to be possible at low (small Explorer mission) cost by using several small, relatively inexpensive satellites in similar orbits to synthesize a large aperture radio telescope above the ionosphere. Historically the opening of a new spectral window has always resulted in major astronomical discoveries, significant insights into astrophysical processes, and an enrichment of our understanding of the universe. The recent IRAS mission is a good example of this.

Groups at JPL, NRL, and other institutions are studying possible missions to map the entire sky with high angular resolution at frequencies between 1-2 MHz and 10-20 MHz. Such low frequencies are totally inaccessible or extremely difficult to observe from the ground. This frequency range is a region essentially unexplored by astronomy, but which is expected to contain a wealth of new information. It is as far removed from the frequency range normally used for radio astronomy as the infrared range surveyed by IRAS is from either the radio or optical spectral regions.

The use of multiple spacecraft will allow individual sources to be imaged with high angular resolution (from $\approx$ a degree at the lowest frequencies to $\approx$ 10 arcseconds at the higher frequencies, limited by scattering in both cases). This will be an extremely valuable improvement over previous single satellite space missions operating at these frequencies (*e.g.*, the Radio As-

© 1988 American Institute of Physics

tronomy Explorers) which had very poor angular resolution. The history of astronomy clearly shows the importance of angular resolution for identifying sources of emission and understanding the physical processes involved. Since angular resolution is proportional to the ratio of the observed wavelength to the telescope diameter, good angular resolution at low frequencies is possible only with very large (multi-kilometer) apertures. A single antenna of this size would be very challenging to build, but an aperture of this size can be synthesized by operating two or more satellites as an interferometer.

SCIENTIFIC GOALS

The objects which can be studied at low frequencies range from solar system objects to distant clusters of galaxies, including galactic objects such as pulsars and supernova remnants and extragalactic objects such as the large radio lobes associtate with many active galaxies. The radio emission from many of these objects peaks at frequencies below 20 MHz, which makes this frequency range unique in its ability to provide information on the total radio luminosities, magnetic field strengths, and radiative lifetimes of these objects. In addition, low frequency measurements of the apparent sizes of pulsars and compact extragalactic radio sources along many lines of sight can tell us about the density irregularities and turbulence in our galaxy's interstellar medium.

Some of the scientific questions to be investigated by a low frequency radio interferometer in space are:

1. What are the effects of scattering and refraction by the interstellar and interplanetary media at low radio frequencies?

2. What is the distribution of low energy cosmic ray electrons? A map of the galactic background non-thermal emission with high resolution and sensitivity will let us investigate this and determine the extent of the Milky Way's halo.

3. What is the galactic distribution of diffuse ionized hydrogen? This can be determined by surveying its absorption effects along lines of sight to a number of discrete galactic and extragalactic sources.

4. What is the emissivity per unit volume in the galaxy? This can be determined in many directions by measuring the foreground emission between the Earth and totally absorbing HII regions.

5. Are there "fossil" radio sources associated with presently radio quiet galaxies and quasars? This would be a sensitive test for earlier epochs of

activity, since at these frequencies the electron lifetimes are a significant fraction of the age of the universe.

6. How do radio source counts differ between very low frequencies and higher frequencies? A survey of the entire sky at 2-20 MHz should detect several thousand sources, sufficient for useful statistical studies.

7. What do the radio spectra of various types of sources look like below 20 MHz? Where do these spectra turn over? Measurements of source spectra will allow us to determine the importance of different emission and absorption processes. The shape of low-frequency spectra also provide information on magnetic field strengths, relativistic paritcle populations, and plasma densities.

8. Is there evidence for coherent radiation processes in galactic and extragalactic objects?

MISSION DESCRIPTION

The technical feasibility of a low-cost mission to survey the entire radio sky and to map individual sources at frequencies between 2 and $\approx$ 20 MHz has been investigated by several groups during the past few years. At JPL, such studies have emphasized simple and low cost spacecraft design. Our major motivation is to maximize the number of spacecraft in the array. This improves the instantaneous coverage of the aperture or Fourier transform (u, v) plane and also improves the ability of self-calibration algorithms to correct the data for instrumental gain variations.

Although short "snapshot" observations may be important for the study of radio wave scattering, the highest quality images will require that data from many orbits be combined. An early JPL study (Kuiper *et al.* 1985, *Radio Science* **20**, 1105) has shown that slow variations in the relative positions of the satellites due to their slightly different orbits will provide a range of array configurations and will result in good (u, v) coverage. This is essential for the formation of accurate high resolution radio images. The maximum useful baseline lengths are $\sim$ 300 km, determined by the expected scattering disk sizes. It should be possible to keep the separations between satellites from exceeding this limit for several months, even with no "station keeping" thrusters on the satellites.

The individual satellites can be very simple, involving only uncooled receivers, nearly omnidirectional (orthogonal dipole) antennas, and a low power telemetry transmitter. They would contain no expendables and no moving parts. Clock stability and synchronization will not be difficult at the long wavelengths considered here, and the bandwidths being received will be quite narrow (≤ 100 kHz) so the telemetry data rates are reasonable.

We plan to send data directly from each satellite to small diameter ground antennas. These would be small enough for their beams to include the entire array ($\sim$ 5 meter diameter at X-band), and might be set up and run by universities. If enough ground stations were built to provide nearly continuous coverage of the array by at least three stations, then a particularly simple method for determining the relative positions of the satellites as a function of time could be used. This involves differential Doppler tracking of the telemetry carriers from each of the satellites.

A number of specific areas are currently being studied, including:

1. Definition of the optimal orbital parameters for the array. Nearly identical circular orbits with altitude $> 30,000$ km are preferred to get above most of the ionosphere and radiation belts.

2. Deployment and initial stabilization of the satellites. For satellites with no thrusters to control the array configuration, it is essential to control the initial deployment carefully. A scheme in which the satellites are deployed from a slowly rotating bus appears feasible, but the bus needs to control the orientation of its spin axis to within a degree.

3. Calibration of data from the array, including the effects of man-made interference. At sufficiently low frequencies the ionosphere provides protection from ground based transmitters, but above the ionospheric plasma frequency we may need to shift or narrow the observing bands in response to strong interfering signals.

4. Techniques to correct data for the time-dependent effects of the extended magnetosphere and solar wind.

CONCLUSIONS

A mission of the type described here would provide useful data on interstellar and interplanetary scattering, in addition to many other areas of astrophysics, at frequencies which have never been explored with high resolution. The mission cost would be small compared to typical space-based scientific missions, and the scientific return would be large. Studies are continuing to refine the satellite design, deployment system, and data analysis requirements.

I am grateful to the members of the NRL Low Frequency Space Array study team, especially Kurt Weiler, for helpful discussions and information on their mission concept. The low-frequency VLBI mission study at JPL has been led by M. Janssen, D. Jones, T. Kuiper, M. Mahoney, and R. Preston. This work was carried out by the Jet Propulsion Laboratory, California Institute of Technology, under contract with the National Aeronautics and Space Administration.

VIII. AGENDAS OF PREVIOUS MEETINGS ON INTERSTELLAR SCATTERING

INTERSTELLAR SCATTERING OF RADIO WAVES

Program of meeting held at the Jet Propulsion Laboratory in Pasadena, CA on 1 February 1974. Organized by D. C. Backer and B. J. Rickett.

Session A - Interstellar Scattering Angle and Pulse Broadening

Introduction - J. J. Broderick (Arecibo Observatory)

Contributions:
- C. C. Counselman, III (MIT) - 'Can thin-screen scattering theory explain the pulse broadening of JP1858 and JP1933?'
- J. M. Rankin (Univ. Iowa) - 'Variations of dispersion and scattering in the Crab nebula pulsar'
- J. J. Broderick (Arecibo) - 'Comparison of pulse broadening in individual and pulse profiles'
- M. H. Cohen (Caltech) - 'Scintillation and apparent angular diameters'
- N. R. Vandenberg (NASA/GSFC) - 'Apparent angular size limits for five pulsars'
- D. C. Backer (NASA/GSFC) - 'Anomalous scattering of the Vela pulsar'
- J. Galt (DRAO) - 'Measurement (or limit) of the scattering angle for PSR 0329+54 at 408 MHz with LBI'
- J. M. Sutton (Univ. Sydney) - 'Angular diameter of PSR 1749-28 from interplanetary scintillation'
- R. L. Mutel (Univ. Col.) - 'Wavelength dependence of interstellar scattering angle'

Session B - Diffraction Pattern and Scintillation

Introduction - D. C. Backer (NASA/GSFC)

Contributions:
- B. J. Rickett (UCSD) - 'Two-station interstellar scattering results'
- J. G. Ables (CSIRO) - 'The Parkes-Ootacamund experiment'
- A. G. Lyne (Jodrell Bank) - 'Three station observations of pulsar scattering diffraction patterns'

Session C - Scattering Theory

Introduction - B. J. Rickett (UCSD)

Contributions:
- R. V. E. Lovelace (Princeton) - 'Radio source position variations arising from a spectrum of interstellar plasma irregularities'
- J. R. Jokipii (Univ. Ariz.) - 'The Markov approximation to strong scintillations in astrophysics: realm of validity and some applications'
- W. M. Cronyn (Clark Lake) - 'Evidence for a power-law dispersion measure structure function from interstellar scattering and dispersion measure observations'
- V. H. Rumsey (UCSD) - 'Scintillation theory'
- W. Newman (Cornell) - 'Multipath distortion and the nature and morphology of pulse broadening'

WORKSHOP ON INTERSTELLAR PROPAGATION OF RADIO WAVES

Program of one-day meeting held at University of California, Berkeley, on 5 September 1986. Organized by D. C. Backer.

Introductory lectures:

- J. Cordes - 'Diffraction Observables and Current Programs'
- R. Narayan - 'Refraction Effects for Galactic and Extragalactic Sources'
- B. Rickett - 'Wave Propagation in the Turbulent ISM: Methods of Analysis'
- J. Higdon - 'Turbulence Wave-Number Spectrum Form and Formation'

Contributions on Refraction:

- R. Frehlich - 'Refractive Intensity Scintillation Under the Influence of and Inner Scale'
- B. Rickett - 'Latitude Dependence of Radio Source Flickering'
- B. Rickett and A. Lyne - 'Refractive Scintillation of Crab Pulsar'
- R. Narayan - 'Refractive Scintillation of 1741-038'
- S. Spangler - 'Analysis of Low-Frequency Variables from Bologna Data'
- J. Cordes - 'Pulsar Slow Intensity Variations'
- R. Narayan - 'Anisotropic Scintillation'

Contributions on High-Resolution Interferometry

- R. Mutel - 'Using Compact Doubles to Text for Refractive Angular Fluctuations'
- D. Backer - 'Interstellar Scattering of SgrA* - Angular Broadening and Limits to Position Wander'
- S. Spangler - 'Angular Broadening of Low-Latitude Sources with VLBI'

Contributions on Diffraction:

- T. Clifton - 'Pulse Broadening in Jodrell Bank 21cm Survey Sample'
- R. Romani - 'Caustics in Pulsar Scintillation'

Contributions on Nature of Micro-Turbulence:

- E. Zweibel - 'Alfven-Wave Propagation in the ISM'
- S. Spangler - 'Production of Density Fluctuations in the Interplanetary Medium'

Author Index

A

Ananthakrishnan, S., 97
Anantharamaiah, K. R., 92, 185
Arons, J., 61

B

Backer, D. C., 111, 228, 229
Bartel, N. H., 106
Bartlett, J. E., 145
Booth, R. S., 195

C

Chernoff, D. F., 174
Clegg, A. W., 174
Clifton, T. R., 180
Coles, W. A., 87, 163
Cordes, J. M., 106, 117, 134, 145, 174, 180, 205, 212, 217
Cornwell, T. J., 92

D

Dennison, B., 97, 150, 195
Dewey, R. J., 217
Diamond, P. J., 195

F

Ferriere, K. M., 70
Fey, A. L., 190
Fiedler, R. L., 97, 150
Foster, R. S., 205
Frehlich, R. G., 169

G

Gibson, C. H., 74
Goodman, J., 200
Gupta, Y., 140
Gwinn, C. R., 106, 129

H

Harmon, J. K., 87
Hewish, A., 82, 150

J

Johnston, K., 150
Jokipii, J. R., 48
Jones, D. L., 222

L

Lestrade, J. F., 122
Lyne, A. G., 140

M

Martinson, A., 195
Max, C. E., 61
Montgomery, D., 60
Moran, J. M., 129
Mutel, R., 106, 122, 190

N

Narayan, R., 17, 92, 185, 200

R

Reid, M. J., 129
Rickett, B. J., 2, 140, 228
Romani, R. W., 156

S

Shull, J. M., 70
Simon, R. S., 97, 150
Simonetti, J. H., 134
Spangler, S. R., 32, 66, 117, 180, 190

W

Weisberg, J. M., 180, 217
Winnberg, A., 195
Wolszczan, A., 106, 145, 212, 217

Z

Zachary, A., 61
Zweibel, E. G., 70

AIP Conference Proceedings

		L.C. Number	ISBN
No. 1	Feedback and Dynamic Control of Plasmas – 1970	70-141596	0-88318-100-2
No. 2	Particles and Fields – 1971 (Rochester)	71-184662	0-88318-101-0
No. 3	Thermal Expansion – 1971 (Corning)	72-76970	0-88318-102-9
No. 4	Superconductivity in d- and f-Band Metals (Rochester, 1971)	74-18879	0-88318-103-7
No. 5	Magnetism and Magnetic Materials – 1971 (2 parts) (Chicago)	59-2468	0-88318-104-5
No. 6	Particle Physics (Irvine, 1971)	72-81239	0-88318-105-3
No. 7	Exploring the History of Nuclear Physics – 1972	72-81883	0-88318-106-1
No. 8	Experimental Meson Spectroscopy –1972	72-88226	0-88318-107-X
No. 9	Cyclotrons – 1972 (Vancouver)	72-92798	0-88318-108-8
No. 10	Magnetism and Magnetic Materials – 1972	72-623469	0-88318-109-6
No. 11	Transport Phenomena – 1973 (Brown University Conference)	73-80682	0-88318-110-X
No. 12	Experiments on High Energy Particle Collisions – 1973 (Vanderbilt Conference)	73-81705	0-88318-111-8
No. 13	π-π Scattering – 1973 (Tallahassee Conference)	73-81704	0-88318-112-6
No. 14	Particles and Fields – 1973 (APS/DPF Berkeley)	73-91923	0-88318-113-4
No. 15	High Energy Collisions – 1973 (Stony Brook)	73-92324	0-88318-114-2
No. 16	Causality and Physical Theories (Wayne State University, 1973)	73-93420	0-88318-115-0
No. 17	Thermal Expansion – 1973 (Lake of the Ozarks)	73-94415	0-88318-116-9
No. 18	Magnetism and Magnetic Materials – 1973 (2 parts) (Boston)	59-2468	0-88318-117-7
No. 19	Physics and the Energy Problem – 1974 (APS Chicago)	73-94416	0-88318-118-5
No. 20	Tetrahedrally Bonded Amorphous Semiconductors (Yorktown Heights, 1974)	74-80145	0-88318-119-3
No. 21	Experimental Meson Spectroscopy – 1974 (Boston)	74-82628	0-88318-120-7
No. 22	Neutrinos – 1974 (Philadelphia)	74-82413	0-88318-121-5
No. 23	Particles and Fields – 1974 (APS/DPF Williamsburg)	74-27575	0-88318-122-3
No. 24	Magnetism and Magnetic Materials – 1974 (20th Annual Conference, San Francisco)	75-2647	0-88318-123-1
No. 25	Efficient Use of Energy (The APS Studies on the Technical Aspects of the More Efficient Use of Energy)	75-18227	0-88318-124-X

No. 26	High-Energy Physics and Nuclear Structure – 1975 (Santa Fe and Los Alamos)	75-26411	0-88318-125-8
No. 27	Topics in Statistical Mechanics and Biophysics: A Memorial to Julius L. Jackson (Wayne State University, 1975)	75-36309	0-88318-126-6
No. 28	Physics and Our World: A Symposium in Honor of Victor F. Weisskopf (M.I.T., 1974)	76-7207	0-88318-127-4
No. 29	Magnetism and Magnetic Materials – 1975 (21st Annual Conference, Philadelphia)	76-10931	0-88318-128-2
No. 30	Particle Searches and Discoveries – 1976 (Vanderbilt Conference)	76-19949	0-88318-129-0
No. 31	Structure and Excitations of Amorphous Solids (Williamsburg, VA, 1976)	76-22279	0-88318-130-4
No. 32	Materials Technology – 1976 (APS New York Meeting)	76-27967	0-88318-131-2
No. 33	Meson-Nuclear Physics – 1976 (Carnegie-Mellon Conference)	76-26811	0-88318-132-0
No. 34	Magnetism and Magnetic Materials – 1976 (Joint MMM-Intermag Conference, Pittsburgh)	76-47106	0-88318-133-9
No. 35	High Energy Physics with Polarized Beams and Targets (Argonne, 1976)	76-50181	0-88318-134-7
No. 36	Momentum Wave Functions – 1976 (Indiana University)	77-82145	0-88318-135-5
No. 37	Weak Interaction Physics – 1977 (Indiana University)	77-83344	0-88318-136-3
No. 38	Workshop on New Directions in Mossbauer Spectroscopy (Argonne, 1977)	77-90635	0-88318-137-1
No. 39	Physics Careers, Employment and Education (Penn State, 1977)	77-94053	0-88318-138-X
No. 40	Electrical Transport and Optical Properties of Inhomogeneous Media (Ohio State University, 1977)	78-54319	0-88318-139-8
No. 41	Nucleon-Nucleon Interactions – 1977 (Vancouver)	78-54249	0-88318-140-1
No. 42	Higher Energy Polarized Proton Beams (Ann Arbor, 1977)	78-55682	0-88318-141-X
No. 43	Particles and Fields – 1977 (APS/DPF, Argonne)	78-55683	0-88318-142-8
No. 44	Future Trends in Superconductive Electronics (Charlottesville, 1978)	77-9240	0-88318-143-6
No. 45	New Results in High Energy Physics – 1978 (Vanderbilt Conference)	78-67196	0-88318-144-4
No. 46	Topics in Nonlinear Dynamics (La Jolla Institute)	78-57870	0-88318-145-2
No. 47	Clustering Aspects of Nuclear Structure and Nuclear Reactions (Winnepeg, 1978)	78-64942	0-88318-146-0
No. 48	Current Trends in the Theory of Fields (Tallahassee, 1978)	78-72948	0-88318-147-9

No. 49	Cosmic Rays and Particle Physics – 1978 (Bartol Conference)	79-50489	0-88318-148-7
No. 50	Laser-Solid Interactions and Laser Processing – 1978 (Boston)	79-51564	0-88318-149-5
No. 51	High Energy Physics with Polarized Beams and Polarized Targets (Argonne, 1978)	79-64565	0-88318-150-9
No. 52	Long-Distance Neutrino Detection – 1978 (C.L. Cowan Memorial Symposium)	79-52078	0-88318-151-7
No. 53	Modulated Structures – 1979 (Kailua Kona, Hawaii)	79-53846	0-88318-152-5
No. 54	Meson-Nuclear Physics – 1979 (Houston)	79-53978	0-88318-153-3
No. 55	Quantum Chromodynamics (La Jolla, 1978)	79-54969	0-88318-154-1
No. 56	Particle Acceleration Mechanisms in Astrophysics (La Jolla, 1979)	79-55844	0-88318-155-X
No. 57	Nonlinear Dynamics and the Beam-Beam Interaction (Brookhaven, 1979)	79-57341	0-88318-156-8
No. 58	Inhomogeneous Superconductors – 1979 (Berkeley Springs, W.V.)	79-57620	0-88318-157-6
No. 59	Particles and Fields – 1979 (APS/DPF Montreal)	80-66631	0-88318-158-4
No. 60	History of the ZGS (Argonne, 1979)	80-67694	0-88318-159-2
No. 61	Aspects of the Kinetics and Dynamics of Surface Reactions (La Jolla Institute, 1979)	80-68004	0-88318-160-6
No. 62	High Energy e^+e^- Interactions (Vanderbilt, 1980)	80-53377	0-88318-161-4
No. 63	Supernovae Spectra (La Jolla, 1980)	80-70019	0-88318-162-2
No. 64	Laboratory EXAFS Facilities – 1980 (Univ. of Washington)	80-70579	0-88318-163-0
No. 65	Optics in Four Dimensions – 1980 (ICO, Ensenada)	80-70771	0-88318-164-9
No. 66	Physics in the Automotive Industry – 1980 (APS/AAPT Topical Conference)	80-70987	0-88318-165-7
No. 67	Experimental Meson Spectroscopy – 1980 (Sixth International Conference, Brookhaven)	80-71123	0-88318-166-5
No. 68	High Energy Physics – 1980 (XX International Conference, Madison)	81-65032	0-88318-167-3
No. 69	Polarization Phenomena in Nuclear Physics – 1980 (Fifth International Symposium, Santa Fe)	81-65107	0-88318-168-1
No. 70	Chemistry and Physics of Coal Utilization – 1980 (APS, Morgantown)	81-65106	0-88318-169-X
No. 71	Group Theory and its Applications in Physics – 1980 (Latin American School of Physics, Mexico City)	81-66132	0-88318-170-3
No. 72	Weak Interactions as a Probe of Unification (Virginia Polytechnic Institute – 1980)	81-67184	0-88318-171-1
No. 73	Tetrahedrally Bonded Amorphous Semiconductors (Carefree, Arizona, 1981)	81-67419	0-88318-172-X

No. 74	Perturbative Quantum Chromodynamics (Tallahassee, 1981)	81-70372	0-88318-173-8
No. 75	Low Energy X-Ray Diagnostics – 1981 (Monterey)	81-69841	0-88318-174-6
No. 76	Nonlinear Properties of Internal Waves (La Jolla Institute, 1981)	81-71062	0-88318-175-4
No. 77	Gamma Ray Transients and Related Astrophysical Phenomena (La Jolla Institute, 1981)	81-71543	0-88318-176-2
No. 78	Shock Waves in Condensed Matter – 1981 (Menlo Park)	82-70014	0-88318-177-0
No. 79	Pion Production and Absorption in Nuclei – 1981 (Indiana University Cyclotron Facility)	82-70678	0-88318-178-9
No. 80	Polarized Proton Ion Sources (Ann Arbor, 1981)	82-71025	0-88318-179-7
No. 81	Particles and Fields –1981: Testing the Standard Model (APS/DPF, Santa Cruz)	82-71156	0-88318-180-0
No. 82	Interpretation of Climate and Photochemical Models, Ozone and Temperature Measurements (La Jolla Institute, 1981)	82-71345	0-88318-181-9
No. 83	The Galactic Center (Cal. Inst. of Tech., 1982)	82-71635	0-88318-182-7
No. 84	Physics in the Steel Industry (APS/AISI, Lehigh University, 1981)	82-72033	0-88318-183-5
No. 85	Proton-Antiproton Collider Physics –1981 (Madison, Wisconsin)	82-72141	0-88318-184-3
No. 86	Momentum Wave Functions – 1982 (Adelaide, Australia)	82-72375	0-88318-185-1
No. 87	Physics of High Energy Particle Accelerators (Fermilab Summer School, 1981)	82-72421	0-88318-186-X
No. 88	Mathematical Methods in Hydrodynamics and Integrability in Dynamical Systems (La Jolla Institute, 1981)	82-72462	0-88318-187-8
No. 89	Neutron Scattering – 1981 (Argonne National Laboratory)	82-73094	0-88318-188-6
No. 90	Laser Techniques for Extreme Ultraviolt Spectroscopy (Boulder, 1982)	82-73205	0-88318-189-4
No. 91	Laser Acceleration of Particles (Los Alamos, 1982)	82-73361	0-88318-190-8
No. 92	The State of Particle Accelerators and High Energy Physics (Fermilab, 1981)	82-73861	0-88318-191-6
No. 93	Novel Results in Particle Physics (Vanderbilt, 1982)	82-73954	0-88318-192-4
No. 94	X-Ray and Atomic Inner-Shell Physics – 1982 (International Conference, U. of Oregon)	82-74075	0-88318-193-2
No. 95	High Energy Spin Physics – 1982 (Brookhaven National Laboratory)	83-70154	0-88318-194-0
No. 96	Science Underground (Los Alamos, 1982)	83-70377	0-88318-195-9

No. 97	The Interaction Between Medium Energy Nucleons in Nuclei – 1982 (Indiana University)	83-70649	0-88318-196-7
No. 98	Particles and Fields – 1982 (APS/DPF University of Maryland)	83-70807	0-88318-197-5
No. 99	Neutrino Mass and Gauge Structure of Weak Interactions (Telemark, 1982)	83-71072	0-88318-198-3
No. 100	Excimer Lasers – 1983 (OSA, Lake Tahoe, Nevada)	83-71437	0-88318-199-1
No. 101	Positron-Electron Pairs in Astrophysics (Goddard Space Flight Center, 1983)	83-71926	0-88318-200-9
No. 102	Intense Medium Energy Sources of Strangeness (UC-Sant Cruz, 1983)	83-72261	0-88318-201-7
No. 103	Quantum Fluids and Solids – 1983 (Sanibel Island, Florida)	83-72440	0-88318-202-5
No. 104	Physics, Technology and the Nuclear Arms Race (APS Baltimore –1983)	83-72533	0-88318-203-3
No. 105	Physics of High Energy Particle Accelerators (SLAC Summer School, 1982)	83-72986	0-88318-304-8
No. 106	Predictability of Fluid Motions (La Jolla Institute, 1983)	83-73641	0-88318-305-6
No. 107	Physics and Chemistry of Porous Media (Schlumberger-Doll Research, 1983)	83-73640	0-88318-306-4
No. 108	The Time Projection Chamber (TRIUMF, Vancouver, 1983)	83-83445	0-88318-307-2
No. 109	Random Walks and Their Applications in the Physical and Biological Sciences (NBS/La Jolla Institute, 1982)	84-70208	0-88318-308-0
No. 110	Hadron Substructure in Nuclear Physics (Indiana University, 1983)	84-70165	0-88318-309-9
No. 111	Production and Neutralization of Negative Ions and Beams (3rd Int'l Symposium, Brookhaven, 1983)	84-70379	0-88318-310-2
No. 112	Particles and Fields – 1983 (APS/DPF, Blacksburg, VA)	84-70378	0-88318-311-0
No. 113	Experimental Meson Spectroscopy – 1983 (Seventh International Conference, Brookhaven)	84-70910	0-88318-312-9
No. 114	Low Energy Tests of Conservation Laws in Particle Physics (Blacksburg, VA, 1983)	84-71157	0-88318-313-7
No. 115	High Energy Transients in Astrophysics (Santa Cruz, CA, 1983)	84-71205	0-88318-314-5
No. 116	Problems in Unification and Supergravity (La Jolla Institute, 1983)	84-71246	0-88318-315-3
No. 117	Polarized Proton Ion Sources (TRIUMF, Vancouver, 1983)	84-71235	0-88318-316-1

No. 118	Free Electron Generation of Extreme Ultraviolet Coherent Radiation (Brookhaven/OSA, 1983)	84-71539	0-88318-317-X
No. 119	Laser Techniques in the Extreme Ultraviolet (OSA, Boulder, Colorado, 1984)	84-72128	0-88318-318-8
No. 120	Optical Effects in Amorphous Semiconductors (Snowbird, Utah, 1984)	84-72419	0-88318-319-6
No. 121	High Energy e^+e^- Interactions (Vanderbilt, 1984)	84-72632	0-88318-320-X
No. 122	The Physics of VLSI (Xerox, Palo Alto, 1984)	84-72729	0-88318-321-8
No. 123	Intersections Between Particle and Nuclear Physics (Steamboat Springs, 1984)	84-72790	0-88318-322-6
No. 124	Neutron-Nucleus Collisions – A Probe of Nuclear Structure (Burr Oak State Park - 1984)	84-73216	0-88318-323-4
No. 125	Capture Gamma-Ray Spectroscopy and Related Topics – 1984 (Internat. Symposium, Knoxville)	84-73303	0-88318-324-2
No. 126	Solar Neutrinos and Neutrino Astronomy (Homestake, 1984)	84-63143	0-88318-325-0
No. 127	Physics of High Energy Particle Accelerators (BNL/SUNY Summer School, 1983)	85-70057	0-88318-326-9
No. 128	Nuclear Physics with Stored, Cooled Beams (McCormick's Creek State Park, Indiana, 1984)	85-71167	0-88318-327-7
No. 129	Radiofrequency Plasma Heating (Sixth Topical Conference, Callaway Gardens, GA, 1985)	85-48027	0-88318-328-5
No. 130	Laser Acceleration of Particles (Malibu, California, 1985)	85-48028	0-88318-329-3
No. 131	Workshop on Polarized ^{3}He Beams and Targets (Princeton, New Jersey, 1984)	85-48026	0-88318-330-7
No. 132	Hadron Spectroscopy–1985 (International Conference, Univ. of Maryland)	85-72537	0-88318-331-5
No. 133	Hadronic Probes and Nuclear Interactions (Arizona State University, 1985)	85-72638	0-88318-332-3
No. 134	The State of High Energy Physics (BNL/SUNY Summer School, 1983)	85-73170	0-88318-333-1
No. 135	Energy Sources: Conservation and Renewables (APS, Washington, DC, 1985)	85-73019	0-88318-334-X
No. 136	Atomic Theory Workshop on Relativistic and QED Effects in Heavy Atoms	85-73790	0-88318-335-8
No. 137	Polymer-Flow Interaction (La Jolla Institute, 1985)	85-73915	0-88318-336-6
No. 138	Frontiers in Electronic Materials and Processing (Houston, TX, 1985)	86-70108	0-88318-337-4
No. 139	High-Current, High-Brightness, and High-Duty Factor Ion Injectors (La Jolla Institute, 1985)	86-70245	0-88318-338-2

No. 140	Boron-Rich Solids (Albuquerque, NM, 1985)	86-70246	0-88318-339-0
No. 141	Gamma-Ray Bursts (Stanford, CA, 1984)	86-70761	0-88318-340-4
No. 142	Nuclear Structure at High Spin, Excitation, and Momentum Transfer (Indiana University, 1985)	86-70837	0-88318-341-2
No. 143	Mexican School of Particles and Fields (Oaxtepec, México, 1984)	86-81187	0-88318-342-0
No. 144	Magnetospheric Phenomena in Astrophysics (Los Alamos, 1984)	86-71149	0-88318-343-9
No. 145	Polarized Beams at SSC & Polarized Antiprotons (Ann Arbor, MI & Bodega Bay, CA, 1985)	86-71343	0-88318-344-7
No. 146	Advances in Laser Science–I (Dallas, TX, 1985)	86-71536	0-88318-345-5
No. 147	Short Wavelength Coherent Radiation: Generation and Applications (Monterey, CA, 1986)	86-71674	0-88318-346-3
No. 148	Space Colonization: Technology and The Liberal Arts (Geneva, NY, 1985)	86-71675	0-88318-347-1
No. 149	Physics and Chemistry of Protective Coatings (Universal City, CA, 1985)	86-72019	0-88318-348-X
No. 150	Intersections Between Particle and Nuclear Physics (Lake Louise, Canada, 1986)	86-72018	0-88318-349-8
No. 151	Neural Networks for Computing (Snowbird, UT, 1986)	86-72481	0-88318-351-X
No. 152	Heavy Ion Inertial Fusion (Washington, DC, 1986)	86-73185	0-88318-352-8
No. 153	Physics of Particle Accelerators (SLAC Summer School, 1985) (Fermilab Summer School, 1984)	87-70103	0-88318-353-6
No. 154	Physics and Chemistry of Porous Media—II (Ridge Field, CT, 1986)	83-73640	0-88318-354-4
No. 155	The Galactic Center: Proceedings of the Symposium Honoring C. H. Townes (Berkeley, CA, 1986)	86-73186	0-88318-355-2
No. 156	Advanced Accelerator Concepts (Madison, WI, 1986)	87-70635	0-88318-358-0
No. 157	Stability of Amorphous Silicon Alloy Materials and Devices (Palo Alto, CA, 1987)	87-70990	0-88318-359-9
No. 158	Production and Neutralization of Negative Ions and Beams (Brookhaven, NY, 1986)	87-71695	0-88318-358-7

No. 159	Applications of Radio-Frequency Power to Plasma: Seventh Topical Conference (Kissimmee, FL, 1987)	87-71812	0-88318-359-5
No. 160	Advances in Laser Science–II (Seattle, WA, 1986)	87-71962	0-88318-360-9
No. 161	Electron Scattering in Nuclear and Particle Science: In Commemoration of the 35th Anniversary of the Lyman-Hanson-Scott Experiment (Urbana, IL, 1986)	87-72403	0-88318-361-7
No. 162	Few-Body Systems and Multiparticle Dynamics (Crystal City, VA, 1987)	87-72594	0-88318-362-5
No. 163	Pion–Nucleus Physics: Future Directions and New Facilities at LAMPF (Los Alamos, NM, 1987)	87-72961	0-88318-363-3
No. 164	Nuclei Far from Stability: Fifth International Conference (Rosseau Lake, ON, 1987)	87-73214	0-88318-364-1
No. 165	Thin Film Processing and Characterization of High-Temperature Superconductors	87-73420	0-88318-365-X
No. 166	Photovoltaic Safety (Denver, CO, 1988)	88-42854	0-88318-366-8
No. 167	Deposition and Growth: Limits for Microelectronics (Anaheim, CA, 1987)	88-71432	0-88318-367-6
No. 168	Atomic Processes in Plasmas (Santa Fe, NM, 1987)	88-71273	0-88318-368-4
No. 169	Modern Physics in America: A Michelson-Morley Centennial Symposium (Cleveland, OH, 1987)	88-71348	0-88318-369-2
No. 170	Nuclear Spectroscopy of Astrophysical Sources (Washington, D.C., 1987)	88-71625	0-88318-370-6
No. 171	Vacuum Design of Advanced and Compact Synchrotron Light Sources (Upton, NY, 1988)	88-71824	0-88318-371-4
No. 172	Advances in Laser Science–III: Proceedings of the International Laser Science Conference (Atlantic City, NJ, 1987)	88-71879	0-88318-372-2
No. 173	Cooperative Networks in Physics Education (Oaxtepec, Mexico 1987)	88-72091	0-88318-373-0

QB 790 .R324 1988

Radio wave scattering in the
 interstellar medium

MAR 2 4 1989